Polymerization Reactors and Processes

J. Neil Henderson, EDITOR
Goodyear Tire and Rubber Company

Thomas C. Bouton, EDITOR
*Firestone Synthetic Rubber
and Latex Company*

Based on a symposium

sponsored by the ACS

Division of Polymer Chemistry

and the American Institute

of Chemical Engineers at

the University of Akron,

Akron, Ohio, October 5–6, 1978.

ACS SYMPOSIUM SERIES **104**

AMERICAN CHEMICAL SOCIETY

WASHINGTON, D. C.   1979

Library of Congress CIP Data

Polymerization reactors and processes.
  (ACS symposium series; 104 ISSN 0097–6156)

  Papers based on a symposium sponsored by the
American Chemical Society and the American Institute
of Chemical Engineers at the University of Akron, Oct.
5-6, 1978.
  Includes bibliographies and index.

  1. Polymers and polymerization—Congresses. 2.
Chemical reactors—Congresses.
  I. Henderson, James Neil, 1924–   II. Bouton,
T. C., 1939–   III. American Chemical Society. IV.
American Institute of Chemical Engineers. V. Series:
American Chemical Society. ACS symposium series;
104.

TP156.P6P63   668   79–12519
ISBN 0–8412–0506–X   ASCMC 8   104 1–407 1979

ACS Symposium Series

Robert F. Gould, *Editor*

FOREWORD

The ACS Symposium Series was founded in 1974 to provide a medium for publishing symposia quickly in book form. The format of the Series parallels that of the continuing Advances in Chemistry Series except that in order to save time the papers are not typeset but are reproduced as they are submitted by the authors in camera-ready form. Papers are reviewed under the supervision of the Editors with the assistance of the Series Advisory Board and are selected to maintain the integrity of the symposia; however, verbatim reproductions of previously published papers are not accepted. Both reviews and reports of research are acceptable since symposia may embrace both types of presentation.

CONTENTS

v

PREFACE

Polymer reactor engineering is an important, evolving branch of technology. This book assembles eighteen papers presented at a joint symposium, "Polymerization Reactors and Processes," sponsored by the American Chemical Society and the American Institute of Chemical Engineers at the University of Akron, October 5 and 6, 1978. The first four papers of the book were plenary lectures at the symposium. The speakers, Gary Poehlein, Joseph A. Biesenberger, A. E. Hamielec, and R. H. M. Simon, were invited to deal at length with selected areas of polymer reactor engineering in which they had made especially significant contributions.

The objective of bringing together practicing engineers in a format suitable for basic instruction as well as for the dissemination of new information was well met both by the formal program and by the informal discussions of the 180 attendees. Considering that the symposium was not part of any larger meeting of the sponsoring societies, the large attendance is a strong indication of the interest in the subject as well as of the quality of the program put together by Irja Piirma and D. C. Chappelear. Credit for the conception and execution of this program must be given to Dr. T. H. Forsyth who saw the need for the meeting and obtained the support of the ACS and AIChE both nationally and locally. Without the umbrella of a larger meeting, much more organization and planning were necessary. In this, Henry Forsyth was aided by a very active committee: Tom Bouton, David C. Chappelear, James Cobb, Joseph N. Feil, Neil Henderson, Thomas A. Kenat, Joginder Lal, Robert W. Lee, Ted Millis, Irja Piirma, Arthur T. Schooley, and Keith C. Williams. In addition to ACS and AIChE, cosponsors were B. F. Goodrich Co., Chemstress Consultants, Firestone Tire & Rubber Co., General Tire & Rubber Co., Goodyear Tire & Rubber Co., PPG Industries, and the Standard Oil Co. (Ohio).

Although the papers represent the whole range of kinds of polymers and processes, there are common themes which reveal the dominant concerns of polymerization reactor engineers. Fully half the papers are concerned rather closely with devising and testing mathematical models which enable process variables to be predicted and controlled very precisely. Such models are increasingly demanded for optimization and computerization of large-scale processes. Another common concern is the improvement of hardware: stirred tanks, tubular reactors, fluidized beds, and RIM molds; continuous, semi-continuous, batch, and precipitation polymerization are all represented. Therefore, taken as a whole, this book represents state-of-the-art polymerization reactor technology.

Finally, the editors wish to thank the authors for their effective oral and written communications and the reviewers for their critical and constructive comments.

Goodyear Tire and Rubber Company   JAMES NEIL HENDERSON
Akron, Ohio 44316

Firestone Synthetic Rubber and Latex Company   THOMAS C. BOUTON
Akron, Ohio 44301
February 21, 1979

Continuous Emulsion Polymerization: Problems in Development of Commercial Processes

GARY POEHLEIN

School of Chemical Engineering, Georgia Institute of Technology, Atlanta, GA 30332

Continuous emulsion polymerization systems are studied to elucidate reaction mechanisms and to generate the knowledge necessary for the development of commercial continuous processes. Problems encountered with the development of continuous reactor systems and some of the ways of dealing with these problems will be discussed in this paper. Those interested in more detailed information on chemical mechanisms and theoretical models should consult the review papers by Ugelstad and Hansen (1), (kinetics and mechanisms) and by Poehlein and Dougherty (2), (continuous emulsion polymerization).

In order to be economically viable, a continuous emulsion polymerization process must be able to produce a latex which satisfies application requirements at high rates without frequent disruptions. Since most latex products are developed in batch equipment, the problems associated with converting to continuous systems can be significant. Making such a change requires an understanding of the differences between batch and continuous reactors and how these differences influence product properties and reactor performance.

Reactor Types:

Before discussing differences between reactors a brief description of reactor types would seem in order. Three classifications are normally recognized: 1. Batch, 2. Semi-Continuous or Semi-Batch, and 3. Continuous. The batch reactor is, in many ways, the simplest. Recipe ingredients are charged and brought to reaction temperature; initiator is then added if it was not part of the original charge; the reaction is carried to the desired degree of conversion and the latex is removed for further processing.

With semi-continuous (more properly, semi-batch) reactors only part of the charge is added at the beginning of the cycle. Usually some reaction time is allowed to pass before the remaining part of the charge is added in a controlled manner. Sometimes

0-8412-0506-x/79/47-104-001$05.00/0

only a portion of the monomer is withheld from the initial
charge while in other cases the secondary feed stream is a
monomer emulsion.

Continuous reactor systems usually consist of stirred tanks
connected in series with all the recipe ingredients fed into the
first reactor and the product removed from the last reactor.
Recipe ingredients can also be added at intermediate points along
the reactor train. Continuous-flow tubular reactors can be used
in series with the tanks, usually as a prereactor in front of
the tanks.

Inhibitor Effect:

Inhibitors can be present in most reaction ingredients. They
are deliberately added to monomers to prevent premature polymerization. Ingredient streams such as monomers are cleaned and
handled carefully to avoid inhibition in fundamental studies,
especially in most academic laboratories. Commercial processes,
however, are usually operated with inhibitors present in the
feed streams, particularly in the monomer. When such ingredients
are used in a batch reactor, a dead time is observed before the
reaction starts.

The simple two-reactor series shown in Figure 1 will be
analyzed to demonstrate the effect of inhibitor on the performance
of continuous systems. Since inhibitor will be present in the
continuously added feed stream, it will serve to reduce the
effective initiation rate in the first reactor. Since inhibitor
is very reactive with free radicals, all inhibitor fed must be
destroyed before significant reaction can take place. Thus the
effective rate of initiation in the first reactor is given by
Equation 1.

$$(1) \quad R_{i,1} = f \left\{ \frac{2K_d[I_2]_o}{1+K_d\Theta_1} - \frac{[H]_o f_H}{\Theta_1} \right\}$$

where $R_{i,1}$ is the net rate of initiation in the first reactor, f
is initiation effectiveness factor, K_d is the initiator decomposition rate constant, $[I_2]_o$ is the initiator concentration in the
mixed feed stream, Θ_1 is the mean residence time in the first
reactor, $[H]_o$ is the inhibitor concentration in the mixed feed
stream, and f_H is the number of free radicals consumed per
inhibitor molecule. Equation 1 is valid only if

$$(2) \quad \frac{2K_d[I_2]_o}{1+K_d\,\Theta_1} \geq \frac{[H]_o f_H}{\Theta_1}$$

In this case the inhibitor concentration in the stream leaving
Reactor 1, $[H]_1$, is zero.

If the inhibitor concentration, $[H]_o$, is larger than
necessary to satisfy the equality of Equation 2 there will be no
polymerization in Reactor 1 and the inhibitor concentration
entering Reactor 2 will be:

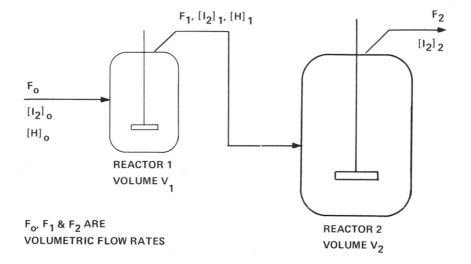

SIMPLIFIED FLOW DIAGRAM —— FRONT
END OF SERIES CSTR SYSTEM

Figure 1.   Continuous flow diagram

(3) $[H]_1 = [H]_o - \dfrac{2K_d[I_2]_o}{(1+K_d\Theta_1)} \quad [\dfrac{\Theta_1}{f_H}]$

In this case the rate of initiation in the second reactor will be given by:

(4) $R_{i,2} = f \left\{ \dfrac{2K_d[I_2]_o}{(1+K_d\Theta_1)(1+K_d\Theta_2)} \right\} - \dfrac{[H]_1 \, f_H}{\Theta_2}$

An examination of the above equations shows that $R_{i,1}$ may be zero, or $R_{i,2}$ may be greater than $R_{i,1}$ even if $R_{i,1}$ is finite. Thus, it may be necessary to add inhibitor to Reactor 2 to slow the reaction so the heat can be removed by the cooling system.

The influence of inhibitor on the performance of a semi-continuous reactor can be, in some ways, similar to both batch and continuous systems. A dead time is usually observed upon addition of the initial charge. When the secondary stream flow is started after some reaction of the initial charge, additional inhibitor flows into the reactor and the initiation rate drops. When this programmed addition is stopped the initiation rate increases; sometimes enough to cause temperature control problems.

Latex Particle Size Distributions:

Particle formation in the early stages of a batch reaction is normally quite rapid. Hence the particle surface area produced is able to adsorb the free emulsifier quite early in the reaction (2 to 10% conversion) and particle formation ceases, or at best slows to a very low rate. Particles formed in the beginning of the reaction would have approximately identical ages at the end of the batch reaction. These particles would be expected to be nearly the same size unless flocculation mechanisms, stochostic differences, or secondary nucleation factors are significant.

The particles in the latex stream leaving a continuous stirred-tank reactor (CSTR) would have a broad distribution of residence times in the reactor. This age distribution, given by Equation 5, comes about because of the rapid mixing of the feed stream with the contents of the stirred reactor.

(5) $A_1(t) = \dfrac{1}{\Theta_1} \, e^{-t/\Theta_1}$

where $A_1(t)$ is the residence time distribution and the particle age distribution in the stream leaving the first tank of the two-tank series shown in Figure 1, and t is time or age.

The residence time distribution for a two-tank system is given by

(6) $A_2(t) = \dfrac{t}{\Theta_1^2} \, e^{-t/\Theta_1}$ if $\Theta_1 = \Theta_2$

or

$$(7) \quad A_2(t) = \frac{e^{-t/\Theta_1} - e^{-t/\Theta_2}}{\Theta_1 - \Theta_2} \quad \text{if } \Theta_1 \neq \Theta_2$$

Graphs of these distributions for various ratios of Θ_1/Θ_2 are shown in Figure 2.

If particle growth rate is known, as a function of particle size, the size distribution can be calculated from Equation 8.

$$(8) \quad U(D) = A(t) \Big/ \left|\frac{dD}{dt}\right|$$

where $U(D)$ is the particle size distribution based on diameter and $\left|dD/dt\right|$ is the absolute value of the rate of diameter change with time. Equation 8 is based on the assumption that particles grow by polymerization rather than flocculation. If Smith-Ewart Case 2 kinetics are followed particle growth in the presence of excess monomer is given by:

$$(9) \quad \frac{dV}{dt} = \frac{\pi D^2}{2} \left(\frac{dD}{dt}\right) = K_1 [M]\bar{n}$$

where K_1 is a constant dependent on polymerization rate constants and swelling parameters, $[M]$ is the monomer concentration at the reaction site and $\bar{n}$ is the time-average number of free radicals per particle ($\bar{n}=0.5$ for S-E Case 2).

When Equation 9 is used in Equation 8 along with the relationships for the residence time distributions one obtains the following dimensionless particle size distributions for one- and two-tank systems.

$$(10) \quad U_1(\mathcal{D}) = 3\mathcal{D}^2 e^{-\mathcal{D}^3}$$

$$(11) \quad U_2(\mathcal{D}) = 3\mathcal{D}^5 e^{-\mathcal{D}^3} \qquad\qquad \text{if } \Theta_1 = \Theta_2$$

$$(12) \quad U_2(\mathcal{D}) = \frac{3\mathcal{D}^2}{m-1} \left(e^{-\mathcal{D}^3/m} - e^{-\mathcal{D}^3}\right) \qquad \text{if } \Theta_2 = m\Theta_1$$

where $\mathcal{D} = D/(6K_1[M]\bar{n}\Theta_1/\pi)^{1/3}$

Equations 11 and 12 are only valid if the volumetric growth rate of particles is the same in both reactors; a condition which would not hold true if the conversion were high or if the temperatures differ. Graphs of these size distributions are shown in Figure 3. They are all broader than the distributions one would expect in latex produced by batch reaction. The particle size distributions shown in Figure 3 are based on the assumption that steady-state particle generation can be achieved in the CSTR systems. Consequences of transients or limit-cycle behavior will be discussed later in this paper.

Semi-continuous reactors can be used to produce very narrow or quite broad particle size distributions depending on the nature of the secondary feed stream and how it is added to the reactor.

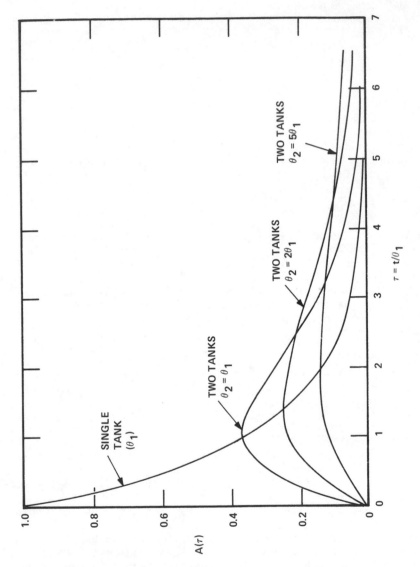

Figure 2. Residence time distributions

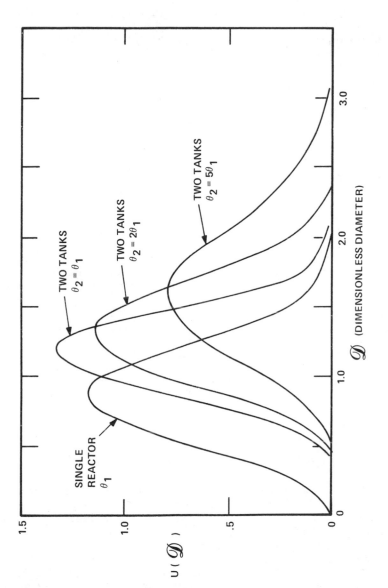

Figure 3. Theoretical particle size distributions

If the secondary feed stream is simply monomer it will not
normally have a major impact on the particle formation reaction
and the particle size distribution can be narrow.

If the secondary stream contains emulsifier it can function
in three ways. When the emulsion feed is started quickly the
added emulsifier can serve to lengthen the particle formation
period and hence to broaden the particle size distribution. When
the emulsion feed is started later and added in such a manner that
the emulsifier is promptly adsorbed on existing particles, one
can obtain quite narrow size distributions. If the emulsion feed
is started later but added rapidly enough to generate free emulsifier in the reaction mixture a second population of particles can
be formed, again yielding a broad size distribution.

Copolymer Composition:

When a batch reactor is used to produce polymer from several
monomers a significant change in copolymer composition can occur
during the course of the polymerization. The first polymer
formed will contain a higher portion of the more reactive monomer
while the final polymer formed will be composed of a larger
fraction of the slow-reacting monomer. More uniform polymer can
be produced by using a semi-continuous system in which a portion
of the more reactive monomer is withheld from the original
charge and added at a carefully programmed rate during the course
of the reaction.

The polymeric material produced in a single stirred-tank
reactor will, except for stochastic variations, be of uniform
composition. This polymer composition can be significantly
different from the composition in the monomer feed mixture unless
the conversion is high. If several tanks are connected in series
the composition of the polymer produced in each reactor can be
quite different. Since most particles are formed in the first
reactor this change in composition in the following reactors can
yield polymer particles in which composition varies with radius
within the particles.

Compositional drift in continuous reactor trains can be altered by introducing feed streams of the more reactive monomer between reactors. This procedure is equivalent to programmed addition of the more reactive monomer in a semi-continuous system.

The proceeding discussion of polymer composition was based on
the assumption that essentially all polymer is formed in the organic phases of the reaction mixture. If a water-soluble monomer,
such as some of the functional monomers, is used, the reactions
taking place in the aqueous phase can contribute to variation in
polymer composition. In fact, in extreme cases, water soluble
polymer can be formed in the aqueous phase. This can happen in
batch, semi-continuous or continuous reactors. The fate of functional monomers could be considerably different among the different reactor types, but detailed studies on this phenomenon have
not been reported.

Reaction Rate:

    Continuous stirred-tank reactors can behave very differently
from batch reactors with regard to the number of particles formed
and polymerization rate. These differences are probably most
extreme for styrene, a monomer which closely follows Smith-Ewart
Case 2 kinetics. Rate and number of particles in a batch reactor
follows the relationship expressed by Equation 13.

(13) $R_p \quad \alpha \quad N \quad \alpha \quad R_i^{0.4} S^{0.6}$

where S is the emulsifier concentration. A single CSTR yields a
different relationship as shown by Equation 14.

(14) $R_p \quad \alpha \quad N \quad \alpha \quad R_i^{0.0} S^{1.0} \Theta^{-.67}$

where Θ is the mean residence time. Equations 13 and 14 represent
rates during interval two in batch polymerization and for intermediate conversions in a CSTR. These two equations illustrate an
important point. That is, even with the same kinetic mechanisms,
the influence of key variables on rate and particle generation
may be quite different between the two reactor types. A summary
of steady-state rates for a number of monomers is given by
Poehlein and Dougherty (2).
    The rate of polymerization with styrene-type monomers is
directly proportional to the number of particles formed. In
batch reactors most of the particles are nucleated early in the
reaction and the number formed depends on the emulsifier available
to stabilize these small particles. In a CSTR operating at
steady-state the rate of nucleation of new particles depends on
the concentration of free emulsifier, i.e. the emulsifier not
adsorbed on other surfaces. Since the average particle size in
a CSTR is larger than the average size at the end of the batch
nucleation period, fewer particles are formed in a CSTR than if
the same recipe were used in a batch reactor. Since rate is
proportional to the number of particles for styrene-type monomers,
the rate per unit volume in a CSTR will be less than the interval-
two rate in a batch reactor. In fact, the maximum CSTR rate will
be about 60 to 70 percent the batch rate for such monomers.
Monomers for which the rate is not as strongly dependent on the
number of particles will display less of a difference between
batch and continuous reactors. Also, continuous reactors with
a particle seed in the feed may be capable of higher rates.
    Reactor production rate depends on average reaction rate and
the fraction of the time the reactor is not operating. With a
batch reactor the reaction rate starts small, increases to a
rather constant value, sometimes increases further to a maximum,
and then decreases rapidly as the monomer concentration falls.
The reaction rate in a continuous reactor is dependent on monomer
conversion but it does not vary with time once steady-state

operation is achieved. This rate can be high for a wide range of
conversions, but it will be low at the high conversion end of the
reactor train. Thus large reactor volumes may be required if
high conversion latexes are to be produced.

Addition of Feed Streams:

Since feed streams are not added after the start of a batch
reaction one need only be concerned with proper initial addition
and blending procedures. Streams flowing into a CSTR, however,
are being introduced into a polymer latex. If added improperly,
these streams can fail to be mixed completely and they can cause
flocculation. Streams should be introduced where they are mixed
rapidly and the ionic concentration should be as low as possible.
Introduction of such streams as initiator solutions at high
concentrations or in the wrong location can cause local
flocculation and/or non-uniform reaction.

Recipe additions can also be important with semi-continuous
reactors. Addition rates influence reactor performance, and
incorrect addition location can lead to non-uniform reaction
within the reactor, localized flocculation, and reactor
short-circuiting.

Unsteady-State Operation:

Achieving steady-state operation in a continuous tank
reactor system can be difficult. Particle nucleation phenomena
and the decrease in termination rate caused by high viscosity
within the particles (gel effect) can contribute to significant
reactor instabilities. Variation in the level of inhibitors in
the feed streams can also cause reactor control problems. Conversion oscillations have been observed with many different monomers.
These oscillations often result from a limit cycle behavior of the
particle nucleation mechanism. Such oscillations are difficult
to tolerate in commercial systems. They can cause uneven
heat loads and significant transients in free emulsifier
concentration thus potentially causing flocculation and the
formation of wall polymer. This problem may be one of the most
difficult to handle in the development of commercial continuous
processes.

One of the most promising ways of dealing with conversion
oscillations is the use of a small-particle latex seed in a feed
stream so that particle nucleation does not occur in the CSTRs.
Berens (3) used a seed produced in another reactor to achieve
stable operation of a continuous PVC reactor. Gonzalez (4) used
a continuous tubular pre-reactor to generate the seed for a
CSTR producing PMMA latex.

Poehlein and Dougherty (2) provide more details on transient
operation problems and some potential control options. Considerable work is currently being conducted in a number of university

and industrial laboratories on approaches to the control of
continuous reactors. These efforts should produce new insights
into this troublesome problem.

Reactor Design:

Ideally one would like a continuous reactor system to
operate indefinitely at the desired steady-state. Unfortunately,
a number of factors can cause shorter runs. Formation of wall
polymer and latex flocculation is one such problem. This phenomenon   can reduce reactor performance (for example, loss of
heat transfer), lower product quality, and shorten run time.

Reactor design can have a significant influence on reactor
performance in a number of ways. Some aspects of reactor design
such as heat transfer, structural design, etc., are reasonably
well-understood. Other phenomena such as mixing details, latex
flocculation, and the formation wall polymer are not completely
understood.

A recent patent (5) describes reactors used for continuous
polychloroprene production which have some interesting features
and claims. These reactors are shown in Figures 4 and 5. They
include the following features:

1.   They are operated completely full thus providing
     no walls in a vapor space which might be a place
     for latex to dry.
2.   The inside surface is smooth with rounded corners
     and no internal fixtures such as baffles.
3.   The axial-flow propellers have been operated with
     a steady flow of 10–15 m^3/min/m^3 reactor volume.
     They have also been operated with oscillating
     motion.
4.   The reactor is completely surrounded by a jacket
     for heating and cooling.
5.   Scale-up is non-geometric with length/diameter
     ratios varying from 2:1 to 30:1. The non-
     geometric scale-up helps to increase heat
     transfer area as reactor volume increases.
6.   The agitator shaft is inclined from 0° to 45°
     with the vertical, and multiple impellers are
     used with longer reactors.

A number of the above features are included to reduce
flocculation and the formation of wall polymer. While fundamental knowledge on flocculation or the formation of wall polymer
is inadequate to establish the effects of all reactor design
variables, the features of the Bayer reactor seem qualitatively
correct. More fundamental work will be necessary to develop an
understanding of the influence of design on reactor performance
and product quality.

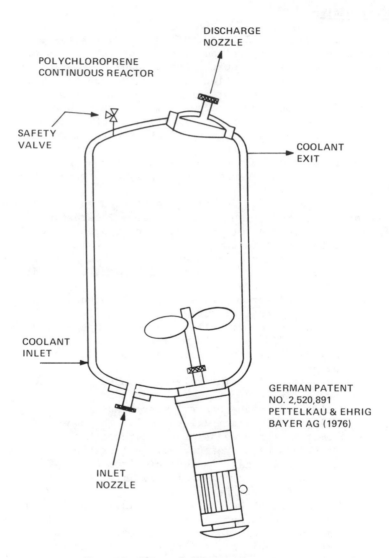

Figure 4.   Short polychloroprene reactor

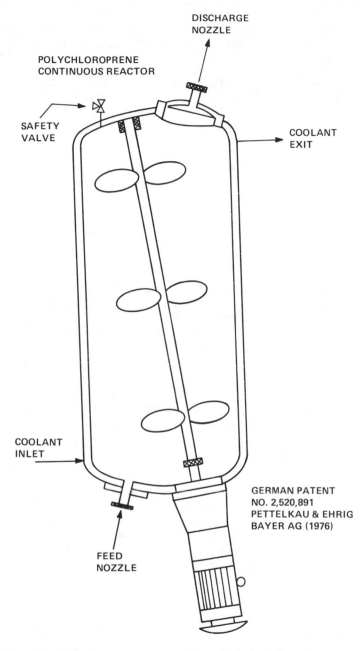

Figure 5.   Polychloroprene reactor with multiple-impeller agitator

Conclusions:

The development of commercial continuous processes involves the consideration of many factors associated with process design and product quality. Most of the factors discussed in this paper will be important. Other, equally significant parameters, may be important for specific polymer products. Failure to deal with any of these problems may mean failure to develop an economical process.

Acknowledgment:

Support from the National Science Foundation (Grants No. GK-36 489 and ENG 75-15 337) is gratefully acknowledged.

Literature Cited:

1. Ugelstad, J. and Hansen, F.K., Rubber Chem. & Technology, (1976), 49(3), 536-609.
2. Poehlein, G.W. and Dougherty, D.J., Rubber Chem. & Technology, (1977), 50(3), 601-638.
3. Berens, A.R., J. Appl. Polym. Sci., 18, (1974), 2379.
4. Gonzalez, P., R.A., M.S. Thesis, Dept. of Chem. Eng., Lehigh University, Bethlehem, Pa. (1974).
5. German Patent No. 2,520,891 (1976), Assigned to Bayer, A.G.

Note: References 1. and 2. contain extensive bibliographies on emulsion polymerization kinetics and continuous emulsion polymerization respectively.

RECEIVED January 19, 1979.

Thermal Runaway in Chain-Addition Polymerizations and Copolymerizations

JOSEPH A. BIESENBERGER

Department of Chemistry and Chemical Engineering, Stevens Institute of Technology, Hoboken, NJ 07030

The objectives of this presentation are to discuss the general behavior of nonisothermal chain-addition polymerizations and copolymerizations and to propose dimensionless criteria for estimating nonisothermal reactor performance, in particular thermal runaway and instability, and its effect upon polymer properties. Most of the results presented are based upon work (1-8), both theoretical and experimental, conducted in the author's laboratories at Stevens Institute of Technology. Analytical methods include a Semenov-type theoretical approach (1,2,9) as well as computer simulations similar to those used by Barkelew (3,4,6,7,10). Analyses of reactor performance are limited to rate functions

$$R = k\pi[c_j]^{n_j} = A\pi[c_j]^{n_j} \exp(-E/R_g T) \tag{1}$$

and thermal energy balances

$$\rho C_p \frac{dT}{dt} = -\Delta HR - \frac{UA}{V}(T - T_R) \tag{2}$$

of the forms shown in equations 1 and 2. Polymer property analyses are limited to chain-addition polymerizations

$$\text{initiation} \quad m_o + m \longrightarrow P_1 \tag{3}$$

$$\text{propagation} \quad P_x + m \xrightarrow{k_p} P_{x+1} \quad x \geq 1 \tag{4}$$

$$\text{termination} \quad P_x \longrightarrow m_x \tag{5}$$

and copolymerizations

$$\text{propagation} \quad P_{x,j} + m_k \xrightarrow{k_{pjk}} P_{x+1,k} \quad \left.\begin{matrix} j \\ k \end{matrix}\right\} = 1,2 \tag{6}$$

with termination, whose general characteristics are shown in
equations 3,4,5 and 6. It should be noted that equations 3 and 5
are written in general form to encompass many different chain
mechanisms and therefore do not necessarily represent elementary
reactions steps. Experimental results quoted herein are limited
to polymerizations and copolymerizations of styrene (S) and acrylonitrile (AN) monomers via free-radical intermediates for which
the following specific reactions obtain. For homopolymerizations
we have

decomposition

$$I \xrightarrow{k_d} 2 P_o \tag{7}$$

initiation

$$P_o + m \xrightarrow{k_i'} P_1 \tag{8}$$

termination

$$P_x + P_y \begin{array}{c} \xrightarrow{k_{tc}} m_{x+y} \\ \xrightarrow{k_{tD}} m_x + m_y \end{array} \tag{9}$$

and for copolymerizations

initiation

$$P_o + m_j \xrightarrow{k_{ij}'} P_{1,j} \qquad j = 1,2 \tag{10}$$

$$- P_j + - P_k \xrightarrow{k_{tjk}} \begin{array}{c} \text{combination} \\ \text{or} \\ \text{disproportionation} \end{array} \left. \begin{array}{c} j \\ k \end{array} \right\} = 1,2 \tag{11}$$

termination

$$- P_iP_j + - P_kP_l \xrightarrow{k_{tijkl}} \begin{array}{c} \text{combination} \\ \text{or} \\ \text{dispropor-} \\ \text{tionation} \end{array} \left. \begin{array}{c} i \\ j \\ k \\ l \end{array} \right\} = 1,2 \tag{12}$$

Termination scheme 11 applies to the geometric mean and phi factor
models and scheme 12 is required for the penultimate effect model.
All the above reaction models were used in attempts to simulate
kinetic data.

Parameters and Variables

Reaction rate functions expressing rate of polymerization R
generally depend upon the molar concentrations of monomer and
initiator, and temperature.

$$R = R([m], [m_o], T) \tag{13}$$

During polymerization, when initiator is introduced continuously following a predetermined feed schedule, or when heat removal is completely controllable so that temperature can be programmed with a predetermined temperature policy, we may regard functions $[m_o(t)]$, or $T(t)$, as reaction parameters. A common special case of $T(t)$ is the isothermal mode, T = constant. In the present analysis, however, we treat only uncontrolled, batch polymerizations in which $[m_o(t)]$ and $T(t)$ are reaction variables, subject to variation in accordance with the conservation laws (balances). Thus, only their initial (feed) values, $[m_o]_o$ and T_o, are true parameters.

In addition to these, we have reactor design parameters: overall heat transfer coefficient U, ratio of reaction volume to heat transfer area $\ell \equiv V/A$ and heat exchange reservoir temperature T_R. While thermodynamic properties $(-\Delta H, \rho C_p)$ and kinetic properties (r, A_p, E_p, A_t, E_t) are determined for the most part by the monomers being polymerized, initiator choice (A_d, E_d) is viewed as a parameter as well as initial monomer concentration $[m]_o$, which can be adjusted through the use of diluents. It will be shown that runaway (R-A) and ignition (IG) phenomena are determined by the values of certain dimensionless groupings, which are made up of the aforementioned parameters. Thus, if R-A is sensitive to one of these groupings, for instance, it will also be sensitive to all other parameters in that grouping.

Frequently function R can be written as a single term having the simple form of equation 1. For instance, with the aid of the long chain approximation (LCA) and the quasi-steady state approximation (QSSA), the rate of monomer conversion, i.e., the rate of polymerization, for many chain-addition polymerizations can be written as

$$R = k_{ap}[m]^P[m_o]^q = A_{ap}[m]^P[m_o]^q \exp(-E_{ap}R_gT) \qquad (14)$$

where k_{ap} is a lumped or composite rate constant. Free-radical homopolymerization is an example $(p = 1, q = 1/2)$ as seen in Table I. Free-radical copolymerization, on the other hand, leads to a sum of terms, each of which is more complex than equation 14, as seen in Table II (Note the presence of function H, given in Table III for various termination modes). To remedy this situation, approximate rate functions for copolymerization of the form of equation 1 are used instead.

In such cases the dimensionless rate function

$$R' \equiv R/(R)_o \qquad (15)$$

can be viewed as a product of separate functions

$$R' = f(t)g(T') \qquad (16)$$

where

TABLE I

RATE FUNCTIONS FOR FREE-RADICAL HOMOPOLYMERIZATIONS

$$R_i = k_i[m_o] = 2f_d k_d[I] \qquad\qquad k_i = fk_d \qquad [m_o] = 2[I]$$

$$R_p \overset{QSSA}{=} k_p[P][m] = k_{ap}[m][I]^{1/2} \qquad k_{ap} \equiv k_p(2fk_d/k_t)^{1/2}$$

$$R_{pt} = k_{pt}[m]^2 \qquad\qquad k_{pt} \equiv k_p^2/k_t$$

$$R = R_i + R_p \overset{LCA}{\approx} R_p$$

$$QSSA \qquad R_i \approx R_t$$

$$LCA \qquad R_p \gg R_i$$

TABLE II

RATE FUNCTIONS FOR FREE-RADICAL COPOLYMERIZATIONS

$$R_{ij} = k_i f_j[m_o] = 2fk_d f_j[I] \qquad j = 1,2$$

$$R = \sum_j R_{ij} = k_i[m_o] = 2fk_d[I] \qquad\qquad \sum_j f_j = 1$$

$$R_{pjk} = k_{pjk}[P_j][m_k] = k_{apjk}[m_j][m_k][I]^{1/2}H \qquad k = 1,2$$

$$k_{apjk} \equiv k_{pjk}k_{p\ell j}(2fk_d/k_{tjj}k_{t\ell\ell})^{1/2} \qquad\qquad \ell = 1,2 \qquad \ell \neq j$$

$$R_{pk} = \sum_j R_{pjk}$$

$$R_p = R_{pk} = \left(\sum_j \sum_k k_{apjk}[m_j][m_k]\right)[I]^{1/2}H$$

$$R = R_i + R_p \approx R_p \qquad \text{from LCA}$$

Symmetry $k_{apjk} = k_{apkj}$ $R_{pjk} = R_{pkj}$ from QSSA

TABLE III

EXPRESSIONS FOR FUNCTION H

Geometric Mean

$$H = \left[\frac{k_{p21}[m_1]}{(k_{t22})^{1/2}} + \frac{k_{p12}[m_2]}{(k_{t11})^{1/2}} \right]^{-1}$$

Phi Factor

$$H = \left\{ \left[\frac{k_{p21}[m_1]}{(k_{t}22)^{1/2}} \right]^2 + 2\phi \frac{k_{p21}k_{p12}[m_1][m_2]}{(k_{t22}k_{t11})^{1/2}} + \left[\frac{k_{p12}[m_2]}{(k_{t11})^{1/2}} \right]^2 \right\}^{-1/2}$$

Penultimate effect

$$H = \left\{ \frac{k_{p21}[m_1]}{(k_{t22})^{1/2}} \left[\frac{r_1[m_1] + \left(\frac{k_{t21}}{k_{t11}}\right)^{1/2}[m_2]}{r_1[m_1] + [m_2]} \right] \right.$$
$$\left. + \frac{k_{p12}[m_2]}{(k_{t11})^{1/2}} \left[\frac{r_2[m_2] + \left(\frac{k_{t12}}{k_{t22}}\right)^{1/2}[m_1]}{r_2[m_2] + [m_1]} \right] \right\}^{-1}$$

where $k_{t11} = k_{t1112}$; $k_{t21} = k_{t2112}$; $k_{t12} = k_{t1221}$; $k_{t22} = k_{t2222}$

$$(R_o) \equiv (k_{ap})_o \; \underset{j}{\pi} \; [C_j]_o^{n_j} \tag{17}$$

$$(k_{ap})_o \equiv A_{ap} \, \exp(-E_{ap}') \tag{18}$$

$$E_{ap}' \equiv E_{ap}/R_g T_o \tag{19}$$

$$f(t) \equiv \pi \, C_j^{n_j} \tag{20}$$

and

$$g(T') \equiv \exp E_{ap}' T'/(1 + T') \tag{21}$$

Frequently it is convenient to write

$$g(\theta) \equiv \exp \theta/(1 + \varepsilon\theta) \tag{22}$$

in lieu of equation 21 where

$$\varepsilon \equiv 1/E_{ap}' \tag{23}$$

We note that under feed conditions $(R')_o = 1$ since $f(0) = 1$ $g(0) = 1$.

Characteristic Times

Balance equations for batch reactors may all be viewed as special cases of the following general equation

$$\frac{dp}{dt} = \sum_j \dot{p}_j \tag{24}$$

where p is an intensive property (molar concentration or temperature) and $\dot{p}_j$ is the rate with which process j causes p to increase in value. When quantities p and $\dot{p}_j$ are made dimensionless through division by their corresponding feed values

$$p' \equiv p/(p)_o \tag{25}$$

$$\dot{p}_j' \equiv \dot{p}_j/(\dot{p}_j)_o \tag{26}$$

the aforementioned balance equations become partly dimensionless, having dimensions of reciprocal time only, and take on the following general form

$$\frac{dp'}{dt} = \sum \lambda_j^{-1} \, \dot{p}_j' \tag{27}$$

in which a characteristic time (CT) for each process may be defined as

$$\lambda_j \equiv (p)_o / (\dot{p}_j)_o \tag{28}$$

We note that equation 15 is an example of equation 26. It can be
shown that all dimensionless parameters, arrived at in the conventional manner by writing equations of type 24 in completely
dimensionless form, can be expressed as quotients of CT's.

Tables IV and V contain appropriate balance equations for
nonisothermal free-radical polymerizations and copolymerizations,
which are seen to conform to equation 24. Following the procedure outlined above, we obtain the CT's for homopolymerizations
listed in Table VI. Corresponding CT's for copolymerizations can
be obtained in a similar way, and indeed the first and fourth
listed in Table VII were. The remaining ones, however, were derived via an alternate route based upon the definitions in Table
VI labeled "equivalent" together with approximate forms for $\dot{p}_j$,
which were necessitated by application of the Semenov-type runaway analysis to copolymerizations, and which will subsequently
be described. Some useful dimensionless parameters defined in
terms of these CT's appear in Tables VIII, IX and X.

Reactor Performance

The condition of thermal runaway (R-A) in polymerization and
copolymerization reactors has been characterized (1,7) by a
rapidly rising temperature $dT/dt \gg 0$ together with an acceleration of the rise $d^2T/dt^2 > 0$. When R-A additionally exhibits
parametric sensitivity it is termed ignition (IG). Beyond its
role as a potential cause of instability, R-A can also affect
conversion efficiency. Specifically, the well-known phenomenon
of dead-ending (D-E), in which conversion of monomer to polymer
is aborted by premature depletion of initiator, is exacerbated by
rising temperatures. This is so because high temperatures accelerate initiator depletion rates much more than monomer conversion rates. The phenomenon can obviously be mitigated by increasing initiator concentration, but this has an adverse effect
on degree of polymerization (DP). The criterion for D-E, shown
in Table XI, was formulated in terms of dimensionless parameter
α_k , shown in Table VIII, which correctly reflects the effects of
feed parameter T_o as well as $[I]_o$, since k_{ax} has a negative
temperature coefficient.

Criteria for R-A and IG, also shown in Table XI, were formulated in terms of dimensionless parameters ε, a, B and b. They
apply to both homopolymerizations and copolymerizations for various initiator systems at or near the condition $T_R = T_o$, and
were developed through modified Semenov-type analyses (1,2,7) and
numerous computer simulations (3,4,6). Owing to the fact that
the dimensionless rate function for homopolymerization contains

TABLE IV

BATCH MATERIAL BALANCE EQUATIONS FOR
FREE-RADICAL POLYMERIZATIONS AND COPOLYMERIZATIONS

Initiator

$$- \frac{d[m_o]}{dt} = - 2 \frac{d[I]}{dt} = R_i/f$$

Monomer

$$- \frac{d[m_k]}{dt} = R_{ik} + R_{pk} \overset{LCA}{\approx} R_{pk}$$

$$- \frac{d[m]}{dt} = R = R_i + R_p \overset{LCA}{\approx} R_p$$

Moment

$$\frac{d[\mu^o]}{dt} = [(2 - r)/2]R_i$$

$$\frac{d[\mu^1]}{dt} = R$$

$$\frac{d[\mu^2]}{dt} = (1 + r)R_i + (3 + 2r)R_p + (2 + r)R_{pt}$$

TABLE V

BATCH ENERGY BALANCE EQUATIONS FOR
FREE-RADICAL POLYMERIZATIONS AND COPOLYMERIZATIONS

Homopolymerizations

$$\rho C_p \frac{dT}{dt} \overset{LCA}{\approx} -\Delta HR_p - (UA/V)(T - T_R)$$

Copolymerizations

$$\rho C_p \frac{dT}{dt} \overset{LCA}{\approx} \sum_j \sum_k (-\Delta H_{jk})R_{pik} - (UA/V)(T - T_R)$$

TABLE VI
CHARACTERISTIC TIMES FOR HOMOPOLYMERIZATIONS

————— Definitions —————

CT	Original	Equivalent	Process
λ_i	$2f[I]_o/(R_i)_o$	$\left(-\dfrac{dm_o}{dt}\right)_o^{-1}$	initiator consumption
λ_m	$[m]_o/(R_p)_o$	$\left(-\dfrac{dm}{dt}\right)_o^{-1}$	monomer conversion
λ_G	$\rho C_p T_o/(-\Delta H)(R)_o$	$\left(\dfrac{G_e}{}\right)_o^{-1}$	heat generation
λ_{ad}	$\rho C_p T_o/(-\Delta H)(R)_o E_{ap}'$	$\left(\dfrac{\partial G_e}{\partial T'}\right)_o^{-1}$	adiabatic induction
λ_R	$\rho C_p \ell/U$	$\left(\dfrac{\partial R_e}{\partial T'}\right)_o^{-1}$	heat removal

TABLE VII
CHARACTERISTIC TIMES FOR COPOLYMERIZATIONS

CT	Original	Equivalent
λ_{jk}	$[m_k]_o/(R_{pjk})_o$	—
Λ_k	$\left(-\dfrac{dm_k}{dt}\right)_o^{-1}$	$\left(\sum_j \lambda_{jk}^{-1}\right)^{-1}$
Λ_m	$\left(-\dfrac{dm}{dt}\right)_o^{-1}$	$\left(\sum_k (x_k)_o \Lambda_k^{-1}\right)^{-1}$
λ_{Gjk}	$C_p T_o/(-\Delta H_{jk})(R_{pjk})_o (H)_o$	—
Λ_G	$(G_e)_o$	$\left(\sum_j \sum_k \lambda_{Gjk}^{-1}\right)^{-1}$
Λ_{ad}	$\left(\dfrac{\partial G_e}{\partial T'}\right)_o^{-1}$	$\sum_j \sum_k \left(E_{apjk}' + \dfrac{\partial H'}{\partial T'}\right)\lambda_{Gjk}^{-1}$

TABLE VIII
DIMENSIONLESS PARAMETERS FOR HOMOPOLYMERIZATIONS

Parameter	Definition	
	For Interpretation	For Evaluation
α_k	$\lambda_m/\lambda_i = B/b$	$1/(k_{ax})_o f[I]_o^{1/2}$
ν_N	—	x_o/α_k
ε	—	$(E'_{ap})^{-1}$
E_{ap}	λ_{ad}/λ_G	$E_{ap}/R_g T_o$
a	λ_{ad}/λ_R	$(U/\ell)T_o/(-\Delta H)(k_{ap})_o E'_{ap}[m]_o[I]_o^{1/2}$
B	λ_m/λ_{ad}	$(-\Delta H)E'_{ap}[m]_o/\rho C_p T_o$
b	λ_i/λ_{ad}	$(-\Delta H)(k_{ax})_o f E'_{ap}[m]_o[I]_o^{1/2}/\rho C_p T_o$

TABLE IX
DIMENSIONAL PARAMETERS FOR COPOLYMERIZATIONS

Parameter	Definition
α_k	Λ_m/λ_i
β_k	Λ_k/Λ_m
$(\nu_1)_o$	$(x_1)_o/\beta_k$
ε	$(E')^{-1}$
E'	Λ_{ad}/Λ_G
a	Λ_{ad}/λ_R
B	Λ_m/Λ_{ad}
b	λ_i/λ_{ad}

TABLE X
EXPRESSIONS FOR H´

Dimensionless Termination Function

$$H' = H/(H)_0$$

Geometric Mean

$$H' = \frac{(x_1)_0(r_1)_0\lambda_1 + (x_2)_0(r_2)_0\lambda_2}{(x_1)_0(r_1)_0\lambda_1 m_1 \exp\frac{E'_{pt21}T'}{2(1+T')} + (x_2)_0(r_2)_0\lambda_2 m_2 \exp\frac{E'_{pt12}T'}{2(1+T')}}$$

Phi Factor

$$H' = \frac{((x_1)_0(r_1)_0\lambda_1)^2 + 2\phi(x_1)_0(r_1)_0\lambda_1 1(x_2)_0(r_2)_0\lambda_2 + ((x_2)_0(r_2)_0\lambda_2)^2}{((x_1)_0(r_1)_0\lambda_1)^2 m_1^2 \exp\frac{E'_{pt21}T'}{2(1+T')} + 2\phi(x_1)_0(r_1)_0\lambda_1(x_2)_0(r_2)_0\lambda_2 m_1 m_2 \exp\frac{(E'_{pt12}+E'_{pt21})T'}{2(1+T')} + ((x_2)_0(r_2)_0\lambda_2)^2 m_2^2 \exp\frac{E'_{pt12}T'}{2(1+T')}}$$

Penultimate Effect

$$H' = \frac{(x_1)_0(r_1)_0\lambda_1\left[\dfrac{(x_1)_0(r_1)_0 + (x_2)_0\Delta_1}{(x_1)_0(r_1)_0 + (x_2)_0}\right] + (x_2)_0(r_2)_0\lambda_2\left[\dfrac{(x_2)_0(r_2)_0+(x_1)_0\Delta_2}{(x_2)_0(r_2)_0+(x_1)_0}\right]}{(x_1)_0(r_1)_0\lambda_1 m_1 \exp\frac{E'_{pt21}T'}{2(1+T')}\left[\dfrac{(x_1)_0(r_1)_0 m_1 \exp\frac{E'_{r1}T'}{(1+T')}+(x_2)_0\Delta_1 m_2 \exp\frac{E'_{D1}T'}{(1+T')}}{(x_1)_0(r_1)_0 m_1 \exp\frac{E'_{r1}T'}{(1+T')}+(x_2)_0 m_2}\right] + (x_2)_0(r_2)_0\lambda_2 m_2 \exp\frac{E'_{pt12}T'}{2(1+T')}\left[\dfrac{(x_2)_0(r_2)_0 m_2 \exp\frac{E'_{r2}T'}{(1+T')}+(x_1)_0\Delta_2 m_1 \exp\frac{E'_{D2}T'}{(1+T')}}{(x_2)_0(r_2)_0 m_2 \exp\frac{E'_{r2}T'}{(1+T')}+(x_1)_0 m_1}\right]}$$

TABLE XI

DIMENSIONLESS CRITERIA FOR REACTOR PERFORMANCE
(polymerizations and Copolymerizations)

Phenomena	Criteria
D-E	$\alpha_k > 1$
R-A	$\varepsilon \ll 1$
	$a < 2$
IG/ERA	$B > 20$
	$b > 100$

only one simple term

$$R_p^\prime = m \, l^{1/2} \exp E_{ap}^\prime T^\prime / (1 + T^\prime) \tag{29}$$

and thus conforms to equation 16, it can be shown by following the procedure leading to equation 27 that the partly dimensionless monomer balance equation

$$-\frac{dm}{dt} = \lambda_m^{-1} R_p^\prime \tag{30}$$

contains only a single function on the RHS and the partly dimensionless energy balance equation takes on a form

$$\frac{dT^\prime}{dt} = \lambda_{G_e}^{-1} R_p^\prime - \lambda_{R_e}^{-1} (T^\prime - T_R^\prime) \tag{31}$$

which is amenable to Semenov-type analysis (1). Pertinent dimensionless parameters were defined in terms of resulting CT's and are listed in Table VIII.

During the development of these criteria the Semenov analysis was extended to systems with heat-exchanger reservoir temperatures different from feed temperatures ($T_R < T_O$) and with delayed runaway (larger value of ε), which resulted in significant concentration drift prior to runaway. Since values of ε for chain-addition polymerizations are not nearly as small as those for the gaseous explosions investigated by Semenov, R-A is not as sensitive nor is it as early in terms of extent of reaction. Thus, the critical value of R-A parameter 'a' is not the same nor is it as clearly defined. Moreover, it is possible to experience insensitive (potentially stable) R-A. Sample experimental results showing sensitive and insensitive R-A have been plotted in Figures 1 and 2, respectively.

In the computer simulations it was necessary to study reaction sequences more complex than those studied by Barkelew, which consequently led to rate functions having double rather than single concentration dependence. Numerous results from both theoretical and computational analyses, including the effects of ε and T_R, have been described elsewhere (see especially Figure 8 of reference 1).

Criteria for sensitivity, B and b, are also criteria for validity of the early R-A approximation (ERA), which says that R-A occurs virtually when $m = l = 1$. While B for most free-radical polymerizations lies within a narrow range, which exceeds the critical value, b varies widely from subcritical to critical values, depending strongly upon choice of initiator and feed parameters $[l]_O$ and T_O. Decreasing values of b generally depress the critical value of 'a' slightly. Computed R-A

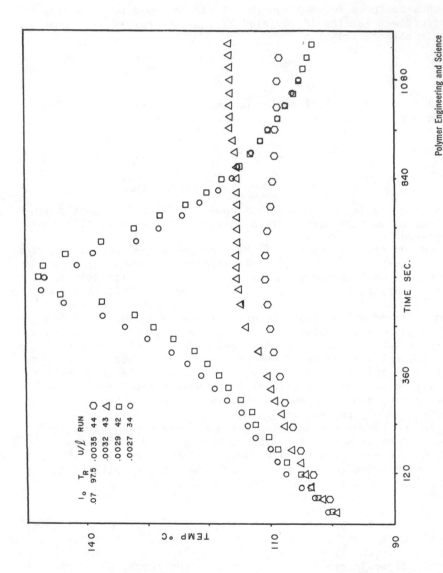

Figure 1. Experimental data from styrene polymerization initiated with benzoyl peroxide showing R-A sensitivity to parameter U/l (5)

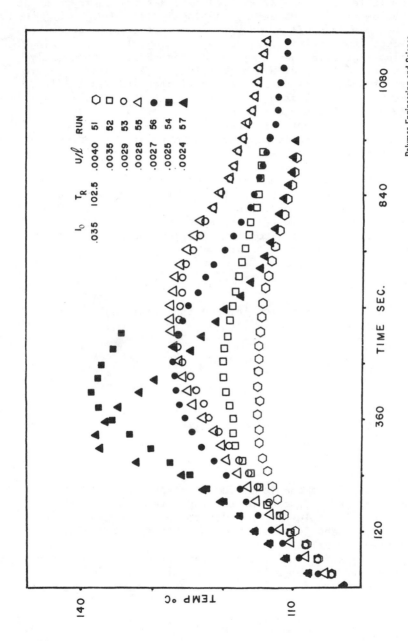

Polymer Engineering and Science

Figure 2.   Experimental data from styrene polymerization initiated with benzoyl peroxide less R-A sensitivity to parameter $[I]_0$ (5)

boundaries for homopolymerizations are shown in Figure 3.

From Tables VIII and XI we find that R-A may be induced ('a' reduced below critical value) by raising T_o, $[I]_o$ or E_{ap} (via E_d) as well as by lowering U/ℓ . We also find that IG may be induced (b increased above critical value) by raising $[I]_o$, lowering T_o or lowering k_d (via lower A_d or higher E_d). Consequently, we must conclude that while a high value of T_o contributes to the onset of R-A, it simultaneously mitigates its sensitivity. Furthermore, while initiators azo-bis-isobutyronitrile ($A_d \sim 10^{15}$ sec^{-1}, $E_d \sim 30$Kcal), benzoyl peroxide ($A_d \sim 10^{13}$ sec^{-1}, $E_d \sim 30$Kcal) and di-tert-butyl peroxide ($A_d \sim 10^{15}$ sec^{-1}, $E_d \sim 37$Kcal) are generally regarded as increasing in "slowness" in the direction listed, because A_d decreases or E_d increases, or both, their value of b increases in the order shown, all other factors remaining equal. Consequently, we must conclude that 'slow' initiators are more likely to produce unstable R-A's than fast ones. The above conclusions involving T_o and initiator choice have been observed experimentally.

The rate function for copolymerization contains a summation of terms, each of which

$$R'_{pjk} = m_j m_k I^{1/2} H' \exp E'_{apjk} T'/(1 + T') \tag{32}$$

is more complex than equation 16, and the resulting monomer balance equation is

$$-\frac{dm}{dt} = \sum_j \sum_k \lambda_{jk}^{-1}(x_k)_o R'_{pjk} \tag{33}$$

The corresponding energy balance

$$\frac{dT'}{dt} = \sum_j \sum_k \lambda_{Gjk}^{-1} R'_{pjk} - \lambda_R^{-1}(T' - T'_R) \tag{34}$$

is consequently not amenable to Semenov-type analysis. The functional form of H' for each of the three termination models cited is given in Table X. In order to remedy this situation, equations 32, 33 and 34 were forced to conform to 29, 30 and 31 by recognizing alternative, equivalent definitions (third column in Table VI) of CT's for homopolymer balances and subsequently applying them to copolymer balances. In this way, approximate copolymer balances

$$-\frac{dm}{dt} = \Lambda_m^{-1} m I^{1/2} \exp E'T'/(1 + T') \tag{35}$$

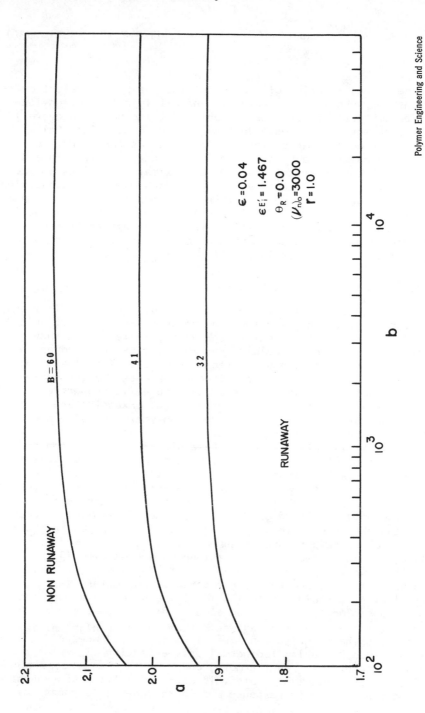

Figure 3. Computed IG boundaries for homopolymerizations (4)

$$\frac{dT'}{dt} = \Lambda_{G_e}^{-1} \, m \, |^{1/2} \, \exp E'T'/(1 + T') - \lambda_{R_e}^{-1}(T - T_R') \qquad (36)$$

were developed in which CT's were obtained using equivalent definitions (second column in Table VII) and all important dimensionless parameters for copolymers (Table IX), including an overall
activation energy E', were defined in accordance with their
homopolymer counterparts (Table VIII). It was found that not
only did parameters a, B and b so defined characterize R-A and IG
behavior of copolymerizations, but approximate equations 35 and
36 closely track the exact balance equations over wide conversion
and temperature ranges, except when one comonomer is exhausted or
when the system is near IG (7). Verification of the applicability
of R-A boundaries to copolymerizations is evident in Figure 4.
Comparisons between "exact" computer models and copolymer approximate forms (CPAF) appear in Figure 5.

Polymer and Copolymer Properties

Owing to the chain nature of chain-addition polymerizations
and copolymerizations with termination, only a small fraction of
the ultimate product molecules grow at any instant, but they grow
to their final size so rapidly that they may be regarded as instantaneous product without appreciable error. The final product
is an accumulation of all instantaneous products formed during
the course of polymerization, and its cumulative properties are
composites of the instantaneous properties. Examples are degree
of polymerization distribution, DPD, copolymer composition distribution, CCD, and their respective average values, DP and CC
(see Table XII). Dispersion of these distributions is consequently the result of the inherent dispersion of the molecular
processes at each instant, termed statistical dispersion, together with the effect of time drift superimposed upon it, termed
drift dispersion, which is a characteristic of batch reactors and
which can only result in greater dispersion if allowed to occur.
Thus, the response of these polymerizations to changes in a parameter, such as temperature or composition, may be viewed as mani-
festing itself in two ways, instantaneous and delayed (3).
It is well known that low values of T_o and $[I]_o$ lead to
high DPs. This is accurately reflected by parameter $(\nu_N)_o$, the
initial kinetic chain length (Table XIII), which is a quotient of
feed composition ratio x_o and dimensionless parameter α_k. Thus,
given x_o, a small value of α_k would seem to favor high initial DP.
On the other hand, criterion $\alpha_k < 1$ signals a downward drift of
instantaneous DP during isothermal polymerization (3) which has
the opposite effect. Furthermore, under nonisothermal conditions,
rising temperatures exacerbate this downward drift. Consequently,
we conclude that drift response and instantaneous response may be

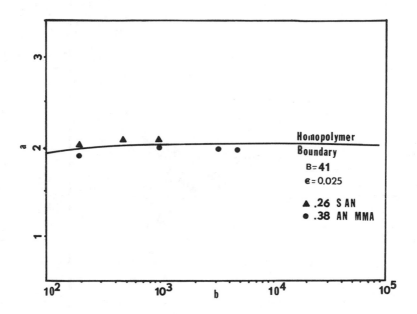

Figure 4.   Computed IG boundaries for copolymerizations

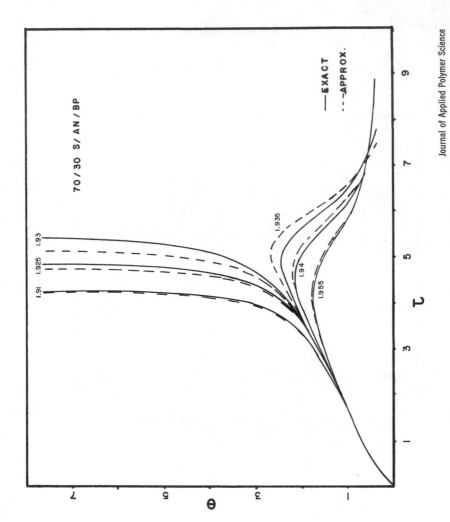

Figure 5. Exact and approximate computer models for SAN copolymerization (2)

TABLE XII

POLYMER AND COPOLYMER PROPERTIES

Instantaneous

$$(\overline{x}_N)_{inst} = 2R/(2 - r)R_i = [2 - r)][1 - k_{ax}[m]/[I]^{1/2}]$$

$$k_{ax} \equiv k_p/(2f_{kd}k_t)^{1/2}$$

$$(\overline{y})_{inst} = R_{p1}/R_p = \nu_1$$

Cumulative

$$\overline{x}_N = [\mu^1]/[\mu^o]$$

$$\overline{x}_w = [\mu^2]/[\mu^1]$$

$$y = ([m_1]_o - [m_1])/([m]_o - [m])$$

TABLE XIII

DIMENSIONLESS CRITERIA FOR POLYMER AND COPOLYMER PROPERTIES

Target Property	Criterion		
	Instantaneous Response	Isothermal Drift Response	Nonisothermal Drift Response
High DP/LCA	large $(\nu_N)_o$	large α_k	large α_k
Narrow DPD	Statistical Dispersion	$\alpha_k = 1$	$\alpha_k > 1$
High cc (comonomer 1)	large $(\nu_1)_o$	large β_k	?
Narrow CCD	Statistical Dispersion	$\beta_k = 1$	?

made to occur in opposite directions by appropriate adjustment of reaction parameters, but it does not necessarily follow that an initial drop in instantaneous DP due to $\alpha_k > 1$, say, can be offset by an expected upward drift in its effect on cumulative DP of the final product. It may be shown (Figure 6), however, that drift can be reduced and total dispersion of DPD thereby kept to a minimum by counteracting downward drift in DP due to R-A temperatures with a large value of α_k . These effects have been summarized in Table XIII.

O'Driscoll, et al. (11) have pointed out that the sign and magnitude of the vertical distance, $\nu_1 - x_1$, between the composition curve and diagonal on the cc diagram are measures of the direction and degree of drift of the instantaneous cc, $(y)_{inst}$ or ν_1 . It can be shown that β_k is an equivalent measure of drift. In fact, the two measures are related via

$$\nu_1 - \eta_1 = (x_1)_o (1 - \beta_k)/\beta_k , \qquad (37)$$

Thus, when $\beta_k < 1$, $(y)_{inst}$ drifts downward and when $\beta_k > 1$ it drifts upward. We therefore conclude that to achieve a target cc high in comonomer 1, say, instantaneous response considerations (Table IX) suggest that, given $(x_1)_o$, a low value for β_k is required, whereas equation 37 indicates that $\beta_k < 1$ would cause the composition to drift downward, opposite to the target direction. Obviously, when $\beta_k = 1$, no drift occurs.

It can be shown that high temperature levels and R-A have virtually no broadening effect on CCD dispersion because β_k has a small temperature coefficient, which frequently even takes on negative values causing drift dispersion to actually lessen at high temperatures. Figures 7 and 8 show the smallness and direction (improvement) of temperature effect on drift, and the ability of β_k to characterize direction (see crossover in Figure 7 and corresponding drifts in Fig. 8) as well as magnitude of drift.

As a final note it should be pointed out that R-A parameters for homopolymerization and copolymerization can be evaluated from initial kinetic rate data using the interpretations given to characteristic times in Tables VI and VII. Coupling between changing reaction viscosity and kinetic constants and other transport properties was neglected because runaway generally occurs early during reaction, and such effects are consequently of minor importance.

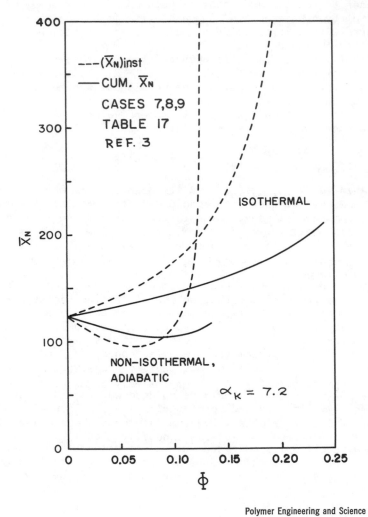

Figure 6. Computed drift curves for instantaneous and cumulative DPs (3)

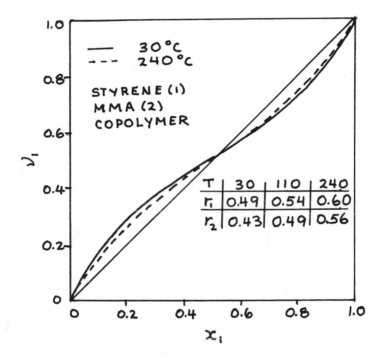

Figure 7.   Computed CC diagram for styrene–methyl methacrylate copolymeri-
zation showing temperature effect on instantaneous CC

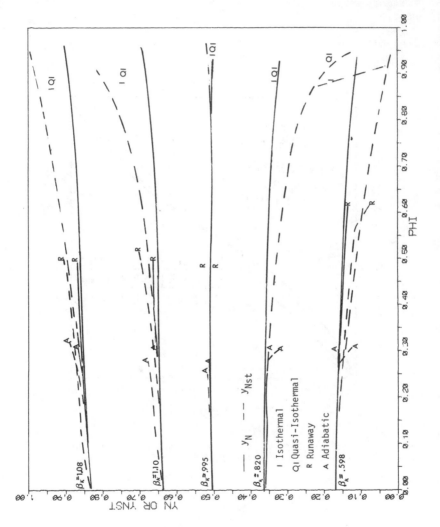

*Figure 8. Computed drift curves for instantaneous and cumulative CCs of styrene–methyl meth-
acrylate polymers initiated with AIBN*

SYMBOLS NOT DEFINED IN TEXT

A = pre-exponential coefficient in rate constant expressions with appropriate subscripts or heat transfer surface area.

c_j = general reaction component j or dimensionless concentration of component j .

E = activation energy in rate constant expressions with appropriate subscripts.

E´ = dimensionless activation energy for copolymerization defined in Table IX.

$E_{rk}' = E_{rk}/R_g T_o$

$E_{Dk}' = (E_{tjkkj} - E_{tkkkk})/R_g T_o$

f = initiator efficiency factor

f_j = initiator efficiency factor for comonomer j

I = initiator or dimensionless initiator concentration

k = rate constant with appropriate subscript

$k_{ap} = k_p (2f k_d/k_t)^{1/2}$

$k_{ax} = k_p/(2f k_d k_t)^{1/2}$

$k_{apjk} = k_{pjk} k_{p\ell j} (2f k_d/k_{tjj} k_{t\ell\ell})^{1/2}$

$k_t = k_{tc} + k_{tD}$

m = monomer or dimensionless monomer concentration

m_j = comonomer j or dimensionless concentration of comonomer j

m_o = generalized initiator or dimensionless concentration of generalized initiator

m_x = x-mer or dimensionless concentration of x-mer

P = active intermediates of all lengths and types

P_j = active j-mer intermediates of all lengths with comonomer j as terminal unit

P_x = active intermediates of all lengths and types

$P_{x,j}$ = active intermediate of length x with comonomer j as terminal unit

P_j = active intermediate of any length with comonomer j as terminal unit

$P_j P_k$ = active intermediate of any length with comonomer j and k as penultimate and ultimate units, respectively

R = rate function for total monomer conversion (rate of polymerization) or any rate function with appropriate subscript

R_{pt} = rate function defined in Table I

R_g = gas constant

r = k_{tc}/k_t or reactivity ratio with appropriate subscript

T´ = $(T - T_o)/T_o$

U = overall heat transfer coefficient

V = reactor volume

$\overline{x}_N$ = number average DP

$\overline{x}_w$ = weight average DP

x_o = $[m]_o/[m_o]_o$

x_j = $[m_j]/[m]$

y = mole fraction of comonomer 1 in copolymer

$\Delta_k = (k_{tjkkj}/k_{tkkkk})_o$

$\phi = k_{t12}/k_{t111}k_{t22}$

$\Phi = 1 - m =$ fraction monomer converted

$\theta = E'_{ap} T'$

$\lambda, \Lambda =$ characteristic times with appropriate subscripts

$\mu^k = k^{th}$ moment of DPD

[ ] = molar concentration

prime$'$ = dimensionless quantity

Subscripts

ap = apparent or lumped
d = decomposition of initiator
G = generation of heat
i = initiation of polymer chains
j,k = comonomer or repeat unit of type j or km where
      $j = 1,2$ and $k = 1,2$
m = monomer depletion
o = feed conditions (except in m_o)
p = propagation
r = reactivity ratio
R = reservoir (thermal) or removal of heat
t = termination
inst = instantaneous

LITERATURE CITED

1. Biesenberger, J. A., Capinpin, R. and Sebastian, D., Appl. Pol. Symp. (1975) 26, 211.
2. Sebastian, D. H. and Biesenberger, J. A., J. Appl. Pol. Sci. (in press).
3. Biesenberger, J. A. and Capinpin, R., Pol. Eng. Sci. (1974) 14, 737.
4. Biesenberger, J. A., Capinpin, R. and Yang, J. C., Pol. Eng. Sci. (1976) 16, 101
5. Sebastian, D. H. and Biesenberger, J. A., Pol. Eng. Sci. (1976) 16, 117.
6. Sebastian, D. H. and Biesenberger, J. A., Pol. Eng. Sci. (in press).
7. Sebastian, D. H. and Biesenberger, J. A., "Chemical Reaction Engineering - Houston", ACS Symp. Series No. 65, Washington, D.C. (1978).
8. Sebastian, D. H., Ph.D. Thesis in Chemical Engineering (1977) Department of Chemistry and Chemical Engineering, Stevens Institute of Technology, Hoboken, New Jersey
9. Frank-Kamenetskii, D. A., "Diffusion and Heat Exchange in Chemical Kinetics", (1955) Princeton University Press, Princeton.
10. Barkelew, C. R., Chem. Eng. Prog. Symp. Ser. (1959), No. 25, 55, 37.
11. O'Driscoll, K. F. and Knorr, R. (1969), Macromolecules 2, 507.

RECEIVED January 15, 1979.

High Conversion Diffusion-Controlled Polymerization

F. L. MARTEN and A. E. HAMIELEC

McMaster University, Hamilton, Canada L8S 4M1

In bulk, solution and emulsion polymerization dramatic physical changes occur during the course of reaction. As polymer concentration increases a point is reached where appreciable chain entanglements occur and eventually a glassy-state transition may result. These physical changes often have a significant effect on both rate of polymerization and molecular weight development and any attempt at modelling such reactions must properly account for these phenomena.

In this manuscript we review the principles of bulk and solution polymerization with particular emphasis on high conversion (high polymer concentrations) rate of polymerization and molecular weight development.

In the literature there is only one serious attempt to develop a detailed mechanistic model of free radical polymerization at high conversions ($\underline{1},\underline{2},\underline{3}$). This model after Cardenas and O'Driscoll is discussed in some detail pointing out its important limitations. The present authors then describe the development of a semi-empirical model based on the free volume theory and show that this model adequately accounts for chain entanglements and glassy-state transition in bulk and solution polymerization of methyl methacrylate over wide ranges of temperature and solvent concentration.

Physical Phenomena of High Conversions

It is appropriate to differentiate between polymerizations occuring at temperatures above and below the glass transition point(T_g) of the polymer being produced. For polymerizations below T_g the diffusion coefficients of even small monomer molecules can fall appreciably and as a consequence even relatively slow reactions involving monomer molecules can become diffusion controlled complicating the mechanism of polymerization even further. For polymerizations above T_g one can reasonably assume that reactions involving small molecules are not diffusion controlled, except perhaps for extremely fast reactions such as those involving termination of small radicals.

0-8412-0506-x/79/47-104-043$07.00/0

Polymerizations Above T_g. Let the polymerization begin in
pure monomer. As the concentration of polymer chains increases
initially one observes a relatively small increase in the termination rate constant. This is related to the effect of polymer concentration on coil size. A reduction in coil size increases the
probability of finding a chain end near the surface and hence
causes an increase in k_t. Soon thereafter at conversions 15-20%
polymer chains begin to entangle causing a dramatic reduction in
radical chain translational mobility giving a rapid drop in k_t.
The onset of chain entanglements depends on polymer concentration,
molecular weight and reaction temperature. It is reasonable to
assume as did Cardenas and O'Driscoll (1,2,3) that larger radical
chains become entangled before smaller ones and that the intrinsic
termination rate of smaller not entangled radicals would be un-
affected until later in the polymerization. The concept that the
termination of some smaller radicals never becomes diffusion-
controlled is however questionable and with the reduction of even
some free volume even these reactions might become diffusion-
controlled.. The significant reduction in termination rate often
causes an almost explosive increase in radical population and rate
of polymerization. The extent of the autoacceleration in R_p depends a great deal upon molecular weight development. For example
in the bulk polymerization of MMA most of the polymer chains are
produced by termination reactions. There is as a consequence a
larger increase in molecular weight as k_t falls and this gives a
multiplier effect in increasing the number of polymer radicals
which are entangled. For polymerization above T_g the propagation
reactions do not become diffusion controlled and as a consequence
a conversion of 100% is approached in a reasonable time scale.

Polymerization Below T_g. For polymerization below T_g the
situation is more complex. To illustrate the phenomena we refer
to Figures 1, 2, 3 and 4. These Figures involve monomers which
are normally polymerized below T_g (MMA, AN, VC). The exception is
polystyrene which is usually polymerized above T_g. Rather than
begin our discussion at low conversion it is convenient, as will
be seen later, to begin at the limiting conversion. When polymerizations are done below T_g the monomer acts as a plasticizer
and a glassy-state occurs at a conversion less than 100%. When
a glass is formed one experiences solid-state polymerization with
a much greater time scale. In the normal time scale the rate of
polymerization may be taken as zero. The existence of this glassy-
state transition has been confirmed for several polymer systems in
bulk and emulsion polymerization by Friis and Hamielec (4) and
more recently by Berens (5) who has independently measured T_g
values for PVC plasticized with its own monomer. This information
is shown in Figures 1, 2 and 3. Figure 1 shows a limiting conversion of 92% for PMMA-MMA at a polymerization temperature of
70°C. In other words a solution of 92% wt PMMA in 8% MMA has a
glass transition point of 70°C. Figure 2 shows limiting conver-

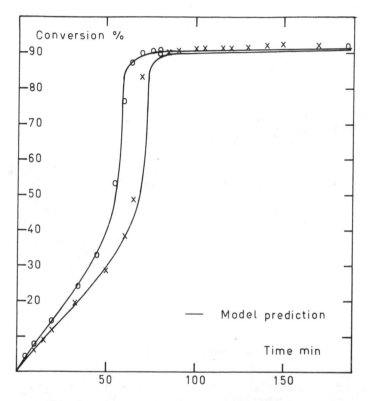

Figure 1. Bulk polymerization of MMA initiated by AIBN (9): temperature 70°C; (○) [I]ₒ = 0.0258 mol/L; (×) [I]ₒ = 0.01548 mol AIBN/L.

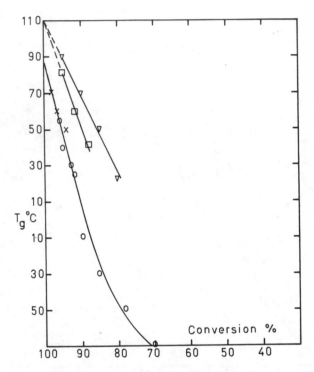

*Figure 2.   Polymerization temperature vs. limiting conversion for different mono-
mer–polymer systems (4): (∇) PMMA; ($\square$) PAN; ($\times$) PS; ($\bigcirc$) PVC.*

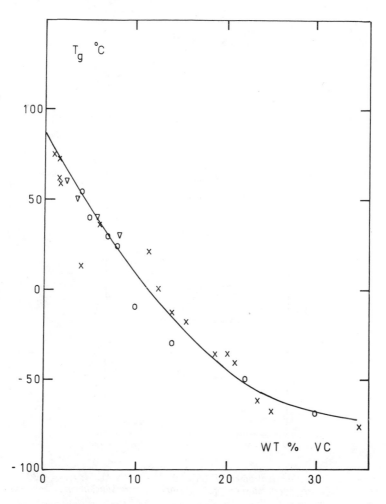

Figure 3. *T_g of PVC vs. content of VC: ($\triangledown$) data measured by means of deviation from Flory–Huggins isotherm (5); ($\times$) data measured thermomechanically (18); ($\bigcirc$) data obtained from limiting conversion (4).*

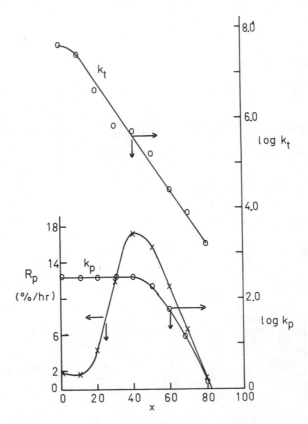

Figure 4.   Bulk polymerization of MMA at 22.5°C with AIBN: $R_I = 8.36 \times 10^{-9}$ mol/L sec. Effect of conversion on propagation and termination rate constants (6).

sions plotted versus polymerization temperature, for PMMA/MMA, PAN/AN, PS/S, PVC/VC systems which have been extrapolated to a limiting conversion of 100% to estimate the T_g of the polymer produced. These T_g values are in general agreement with values measured by DSC and mechanical spectroscopy. Later in the manuscript it will be shown that the effect of residual monomer on glass transition point as measured via kinetics and limiting conversion, agrees with the free-volume theory. From the observation of limiting conversions below 100%, it is clear that even relatively slow propagation reactions involving the small monomer molecule become diffusion controlled well below the limiting conversion. This is confirmed by measurements of k_p after Hayden and Melville ([6]), in Figure 4 where a limiting conversion to about 80% was observed and it was found that k_p already began to drop in value at a conversion of about 50%. These observations have significant implications as far as termination reactions are concerned. It must be concluded that the magnitude of k_t even for the smallest radical reactions must be diffusion-controlled probably from the onset of chain entanglements. In other words k_t for termination of small radicals must decrease significantly with conversion.

As mentioned above for bulk MMA polymerization, polymer chains are produced mainly by termination reactions and hence the variation of k_t, R_p and molecular weight are strongly coupled phenomena. This is particularly true from the onset of chain entanglements but later in the polymerization when k_t has fallen appreciable transfer to monomer becomes an important polymer producing reaction, limiting the ultimate molecular weights that can be obtained. In certain polymerizations such as VC and styrene above 100°C, transfer reactions control molecular weight development and the autoacceleration in R_p is smaller with virtually no effect on molecular weight developments.

Desirable Features of a Polymerization Model at High Conversion.

A useful model should account for a reduction of k_t and k_p with increase in polymer molecular weight and concentration and decrease in solvent concentration at polymerization temperatures both below and above the T_g of the polymer produced. For a mechanistic model this would involve many complex steps and a large number of adjustable parameters. It appears that the only realistic solution is to develop a semi-empirical model. In this context the free-volume theory appears to be a good starting point.

Mechanistic Model - Cardenas and O'Driscoll ([1], [2], [3]).

The basis of this model involves the assumption that radicals with a chain length $> x_c$ are entangled with the following relationship based on viscosity measurements used to establish x_c.

$$K_c = \phi_p \cdot \bar{x}_c^\beta \qquad (1)$$

where ϕ_p is the polymer volume fraction.
           x_c is the number average chain length at the point of
           chain entanglement.
           β is an adjustable parameter (β usually unit for vis-
           cosity measurements).
           K_c is the entanglement constant.

In applying equation (1) Cardenas and O'Driscoll use x_c as
the critical chain length for chain entanglement and permit x_c to
decrease as ϕ_p increases during the polymerization according to
equation (1). Therefore, during the course of polymerization they
note three kinds of termination reactions:

$$R^{\cdot}_r \; + \; R^{\cdot}_s \; \xrightarrow{k_t} \qquad \text{where } r, \; s < x_c$$

$$R^{\cdot}_r \; + \; (R_s^{\cdot})_e \; \xrightarrow{k_{t_c}} \qquad \text{where } r < x_c, \; s > x_c$$

$$(R^{\cdot}_r)_e \; + \; (R_s^{\cdot})_e \; \xrightarrow{k_{t_e}} \qquad \text{where } r > x_c, \; s > x_c$$

It is assumed that k_t is independent of conversion and that
k_{t_e} the termination constant for entangled radicals is given by

$$\alpha \; = \; \left(\frac{k_{t_e}}{k_t}\right)^{\frac{1}{2}} \; = \; \alpha_o \left(\frac{K_c}{\phi_p \, x_n^{\beta}}\right)^{\frac{1}{2}} \tag{2}$$

and finally

$$k_{t_c} \; = \; (k_t \, k_{t_e})^{\frac{1}{2}} \tag{3}$$

This model accounts for the coupling between molecular weight development and autoacceleration in R_p. However, two of the basic
assumptions
     i) k_t is independent of conversion
     ii) k_p is independent of conversion
are certainly not valid for polymerizations below T_g. This model
does not account for a glassy state-transition and hence cannot
predict the observed limiting conversion. For temperatures above
T_g it may prove to be successful. Unfortunately, it has not yet
been evaluated under these conditions.

A Model Based on Free-Volume Theory.

Three main problems are involved in model development.

1. Determination of the conversion at which significant chain entanglements first occur.
2. Development of a relationship which gives the decrease in the termination rate constant as a function of temperature and polymer molecular weight and concentration.
3. Development of a relationship which gives the decrease in the propagation rate constant as a function of temperature and polymer molecular weight and concentration.

The rate of polymerization can be shown to be in the case of isothermal bulk polymerization

$$\frac{dx}{dt} = (\frac{k_p^2}{k_t})^{\frac{1}{2}} \sqrt{\frac{f \cdot k_d [I]_o}{(1 - \varepsilon x)}} (1 - x) \exp(-\frac{k_d \cdot t}{2}) \qquad (4)$$

where k_p = propagation rate constant.
k_t = termination rate constant.
f = initiator efficiency.
k_d = decomposition constant of initiator.
$[I]_o$ = initial initiator concentration.
ε = $(d_p - d_M)/d_p$, volume contraction factor.
d_p = density of polymer.
d_M = density of monomer.
x = degree of conversion.
t = time

In order to estimate the dependence of the termination rate constant on conversion, molecular weight and temperature, the following is assumed: k_t becomes diffusion controlled when the diffusion coefficient for a polymer radical D_p becomes less than or equal to a critical diffusion coefficient $\bar{D}_{p_{cr}}$

$$D_p \leq D_{p_{cr}} \qquad (5)$$

It is further assumed that the termination rate constant beyond this conversion can be expressed by eq. (6a) and at the critical point (6b).

$$k_t = k_1 D_p \qquad (6a) \qquad \qquad k_{t_{cr}} = k_1 D_{p_{cr}} \qquad (6b)$$

where k_1 = temperature dependent proportionality constant.
D_p = diffusion coefficient of polymer radical.
$D_{p_{cr}}$ = critical diffusion coefficient of polymer radical.

If no entanglements are present, the diffusion coefficient of a polymer molecule is, according to Beuche (7), given as

$$D_P = (\phi_o \delta^2 /k_2 \cdot M) \exp (-A/V_F) \qquad (7)$$

where M = molecular weight of polymer (monodispersed).
ϕ_o = jump frequency.
δ = jump distance.
k_2 = constant
A = constant
V_F = free volume.

V_F in the case of bulk or solution polymerization is equal to

$$V_F = (0.025 + \alpha_P(T - T_{g_P})) \frac{V_P}{V_T} + (0.025 + \alpha_M(T - T_{g_M})) \tag{8}$$

$$\frac{V_M}{V_T} + (0.025 + \alpha_S(T - T_{g_S})) \frac{V_S}{V_T}$$

where

M, P and S denote monomer, polymer and solvent respectively.
T = polymerization temperature.
V = volume. V_T = total volume.
T_g = glass transition point of monomer.
α = $\alpha_\ell - \alpha_g$
α_ℓ = expansion coefficient for the liquid state.
α_g = expansion coefficient for the glassy state.

It is further established that

$$T_{g_P} = T_{g_\infty} - \frac{Q}{\bar{M}_N} \tag{9}$$

where T_{g_∞} is the glass temperature of the infinite molecular weight polymer and $\bar{M}_n$ is the cumulative number average molecular weight. Q is a constant independent of temperature.

If equation (6a) is inserted into (7) one obtains

$$k_t = k_1 (\phi_o \cdot \delta^2 / k_2 M) \exp(-A/V_F) \tag{10}$$

For a polymer with a molecular weight distribution the proper molecular weight average to use in equations (7) and (10) can be determined using the following considerations. In the case of a heterogeneous polymer it has been shown that (7,19,20).

$$\eta = k_4 \cdot \bar{M}_w$$

for not entangled polymer solutions, and

$$\eta = k_4 \bar{M}_w^{3.5}$$

for entangled polymer solutions.
The diffusion coefficients of entangled polymers in solution will most certainly depend on the viscosity of the medium and vice versa. It is reasonable therefore to expect that the diffusion coefficient would correlate well with the weight average molecular weight of the polymer. $\bar{M}_w$ is therefore used with equation (10) giving

$$k_1 (\phi_o \cdot \delta^2 / k_2 \bar{M}_w^m) \exp(-A/V_F) = k_t \tag{10a}$$

for unentangled polymer solutions

$$\tilde{k} \; (\phi_o \cdot \delta^2 / \tilde{k}_2 \bar{M}_w^{\;n}) \; \exp(-A/V_F) = k_t \tag{10b}$$

for entangled polymer solutions.

If equation (10a) is combined with (6b) and rearranged, one has

$$K_3 = (\frac{\Psi_1}{k_{t_{cr}}}) = \bar{M}_{wcrl}^{\;m} \exp(+A/V_{F_{crl}}) \tag{10c}$$

$\bar{M}_{wcrl}$ and $V_{F_{crl}}$ must be estimated for each polymerization by satisfying equation (10c). K_3 depends only on temperature and A is a constant and therefore the relationship between $\bar{M}_{wcrl}$ and $V_{F_{crl}}$ depends on temperature alone through equation (10c) At a constant temperature, the magnitude of both $\bar{M}_{wcrl}$ and $V_{F_{crl}}$ can change with initiation rate or concentration of solvent or chain transfer agent. The relationship given by equation (10c) is however the same.

If it is assumed that chain enganglements occur soon after k_t becomes diffusion controlled, then one has as a good approximation

$$k_t = (\frac{\Psi_2}{\bar{M}_w^{\;n}}) \; \exp(-A/V_F) \tag{10d}$$

$$k_{t_o} = k_{t_{cr}} = (\frac{\Psi_2}{\bar{M}_{wcrl}^{\;n}}) \; \exp(-A/V_{F_{crl}}) \tag{11}$$

Combining equations (10d) and (11), k_t is obtained as a function of conversion and the weight average molecular weight

$$(\frac{k_t}{k_{t_o}}) = (\frac{\bar{M}_{wcrl}}{\bar{M}_w})^n \; \exp(-A(\frac{1}{V_F} - \frac{1}{V_{F_{crl}}})) \tag{12}$$

While K_3 and A as explained later were estimated using a fit to experimental data m and n arbitrarily set equal to 0.5 and 1.75 respectively.

The remaining problem in the model development is to estimate the decrease in k_p as a function of conversion. As the reaction proceeds beyond the point of chain entanglement, a critical conversion is reached where the propagation reaction becomes diffusion controlled and k_p begins to fall with further increase in polymer concentration. At the critical conversion, one may write

$$\psi_3 \, D_{M_{cr}} = k_{p_0} \tag{13}$$

where k_{p_0} = the propagation constant below the critical conversion.

$D_{M_{cr}}$ = the diffusion coefficient of the monomer at the critical conversion.

ψ_3 = a proportionality factor.

Beuche (7) gives the following expression for the diffusion coefficient of a small molecule in a polymer solution. This equation also known as the Dolittle equation is

$$D_M = (\phi_2 \, \delta_2^2/6) \exp(-\,^B/V_F) \tag{14}$$

Beyond the critical conversion k_p is given by

$$k_p = \psi_3 \exp(-\,^B/V_F) \tag{15}$$

and

$$\frac{k_p}{k_{p_0}} = \exp\left(-B\left(\frac{1}{V_F} - \frac{1}{V_{Fcr2}}\right)\right) \tag{16}$$

According to Beuche (7) B = 1.0 and this value is used here.

The general rate expression for the complete conversion interval is

$$\frac{dx}{dt} = \left(\frac{k_{p_0}}{k_{t_0}^{\frac{1}{2}}}\right)\left(\frac{\bar{M}_w}{\bar{M}_{wcr1}}\right)^\alpha \left(\exp\left(-B\left(\frac{1}{V_F} - \frac{1}{V_{Fcr2}}\right)\right)\right)\left(\exp\left(\frac{A}{2}\left(\frac{1}{V_F} - \frac{1}{V_{Fcr1}}\right)\right)\right)$$

$$\left(\frac{fk_d[I]_0}{(1 - \epsilon x)}\right)^{\frac{1}{2}} (1 - x) \exp(-\,^{k_d t}/2) \tag{17}$$

Conversion Interval 1 : $\alpha = 0,$ $B = 0,$ $A = 0$
            Interval 2 : $\alpha = 0.875,$ $B = 0,$ $A = 1.11$
            Interval 3 : $\alpha = 0.875,$ $B = 1.0,$ $A = 1.11$

The determination of the conversion intervals are discussed later, after the molecular weight development.

The instantaneous number and weight average degrees of polymerization are given by

$$\tau = \frac{2}{\bar{X}_W} = \frac{1}{\bar{X}_N} = \frac{k_t \, R_p}{k_p^2 [M]^2} + C_M + C_S \frac{[S]}{[M]}$$

when termination is solely by disproportionation; and the cumulative averages by

$$\text{cum } \bar{M}_N = \frac{xM_0}{\int_o^x \tau dx} \tag{18}$$

$$\text{cum } \bar{M}_W = \frac{2M_o}{x} \int_o^x \frac{dx}{\tau} \tag{19}$$

It should be understood that the weight average molecular weights appearing in equation (17) are cumulative ones.

The conversion-time history is obtained by simultaneous solution of equations (17) and (19).

The conversion intervals are determined in the following way: Values of

$$A \text{ and } K_3$$

is guessed and equations (17) and (19) are integrated in interval 1. The calculated cum $\bar{M}_W$ and a calculated V_F are substituted into equation (10c). The end of interval 1 is reached when the equation is satisfied. The integration is carried further into interval 2 with the appropriate parameters. The error of fit in these intervals is noted and

$$A \text{ and } K_3$$

adjusted accordingly. This procedure thus establishes the correct end of interval 1. We next guess the critical free volume where k_p begins to fall and then integrate through interval 3 to limiting conversion. The error of fit is used to establish $V_{F_{cr2}}$ and the critical conversion.

Comparison of Simulated and Measured Rate Data.

Model Parameters used with Equations (17) and (19) for MMA Polymerization.

$\dfrac{k_{p_o}}{k_{t_o}^{\frac{1}{2}}}$: Data after Balke(9), Ito(10) and Hayden and Melville(6) were correlated with an Arrhenius type plot giving the following equation which was used in all the simulations.

$$\frac{k_{p_o}}{k_{t_o}^{\frac{1}{2}}} = 4.48 \cdot 10^2 \exp(\frac{-4.1 \text{ kcal/mol}}{T \cdot R}) \text{ (1/mole min)} \tag{20}$$

$\bar{M}_{wrc_1}$ and $V_{F_{cr1}}$: Refer to equation (10c) and equation (21) which follows.

$$K_3 = \frac{\psi_1}{k_{t_c}} = 0.563 \exp(8,900 \text{ cal/mol/RT}) \tag{21}$$

where
$$R = 1.986 \text{ cal/(g mole)}(°K)$$

```
T   in   (°K)
A   =    1.11
m   =    0.5
```

Equations (17) and (19) are integrated in Interval 1 until the cumulative $\bar{M}_W$ and V_F satisfy equation (10c). This provides $\bar{M}_{wcr1}$ and $V_{F_{cr1}}$. Refer to Figure 5 for the actual K_3 data.

$$V_{F_{cr2}} = 0.066 \tag{22}$$

with B = 1. This parameter is independent of temperature and polymer molecular weight and concentration.

V_F :   The free volume is calculated using equations (8) and (9) with the following parameters.

$$\alpha_P = 0.48 \times 10^{-3} \quad (°C)^{-1} \quad (11)$$
$$T_g = 114 \quad\quad\quad\quad (°C) \quad\quad (11)$$
$$\alpha_M = 10^{-3} \quad\quad\quad\quad (°C)^{-1} \quad (11)$$
$$T_{g_M} = -106 \quad\quad\quad\quad (°C) \quad\quad (11)$$
$$\alpha_S = 10^{-3} \quad)$$
$$T_{g_S} = -102°C \quad) \quad \text{benzene as solvent}$$

$$Q = 2.208 \cdot 10^5 \; [\text{g/mole}] \text{degree}(11)$$
$$d_M = (0.973 - 1.164 \cdot 10^{-3} \cdot t°C)\text{g/cm}^3$$
$$d_P = 1.2 \text{g/cm}^3 \quad\quad\quad\quad (12)$$

It should be noted that polymer volume fraction is readily converted to conversion.

k_d :   AIBN (13) $k_d = 6.32 \cdot 10^{16} \exp(\dfrac{-15.46 \text{ kcal/mole}}{T})\text{min}^{-1}$

$$\dots \;\; (23a)$$

BOP (14)= $k_{d_{70}} = 7.5 \cdot 10^{-4} \text{min}^{-1}$, $k_{d_{50}} = 3.64 \cdot 10^{-5} \text{ min}^{-1}$

f was set equal to unity in all the simulations.

ε : ε the volume contraction factor is calculated using density data as

$$\varepsilon = \frac{d_P - d_M}{d_P} \tag{24}$$

C_M : $C_M = 8.93 \cdot 10^{-4} \exp(\dfrac{-2.24 \text{ kcal/mole}}{R \cdot T}) \tag{25}$

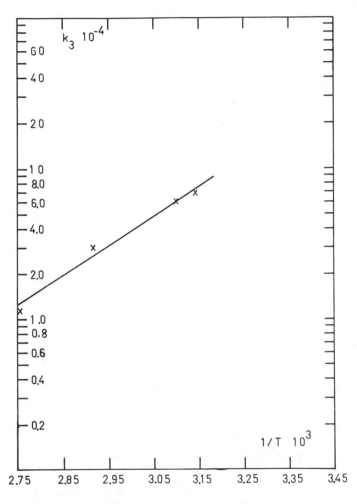

Figure 5. Arrhenius plot of k_3

Bulk Polymerization of MMA

Balke's Data. Figures 1, 6, 7 and 8 show rate data after Balke in the temperature range, 50 - 90°C. Also included is one rate curve after Nishimura (15). There is obviously excellent agreement even at limiting conversion. Figures 9 and 10 show a comparison of measured and predicted weight average molecular weights. The agreement with $\bar{M}_N$ is excellent, but with $\bar{M}_W$ only fair at intermediate conversions near the onset of chain entanglements. The reproducibility of $\bar{M}_W$ when a high molecular weight spike is generated is rather poor and perhaps this may explain some of the deviation.

Ito's Data. Figure 11 shows Ito's(10) rate data at 45°C for a very wide range of initiator concentrations (AIBN: 0.2-0.00625 gmole/ℓ). The agreement is excellent showing that the large changes in molecular weights can be accounted for in our model.

Solution Polymerization of MMA

Schulz's Data (16). Figures 12 and 13 show excellent agreement between simulated and measured rates for a wide range of solvent concentrations (benzene: 0-0.927 liter benzene to 0.103 liter MMA and benzoyl peroxide: 0.0413 gmole/ℓ at 50°C and 70°C). No doubt measured and predicted molecular weights would have been in good agreement. Transfer to benzene was neglected in the simulations.

It should be mentioned that the predicted curve at highest benzene level in Figure 13 agrees with classical kinetics (no diffusion-control). It is not clear therefore why measured data at even higher benzene concentrations do not agree with classical kinetics. There may be some subtle chemical interactions at these high solvent levels. Duerksen(17) found similar effects with styrene polymerization in benzene and had to correct k_p for solvent.

Conclusions

A new rate model for free radical homopolymerization which accounts for diffusion-controlled termination and propagation, and which gives a limiting conversion, has been developed based on free-volume theory concepts. The model gives excellent agreement with measured rate data for bulk and solution polymerization of MMA over wide ranges of temperature and initiator and solvent concentrations.

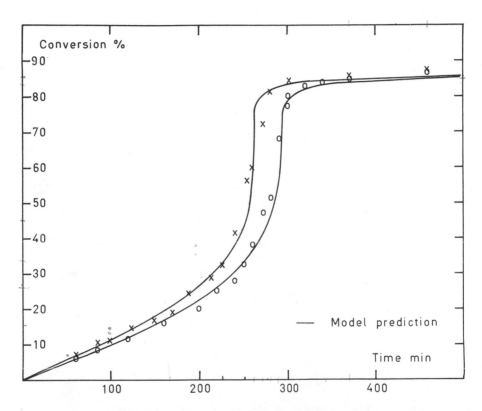

Figure 6.   Bulk polymerization of MMA at 50°C: (✗) [I]$_o$ = 0.02018 mol AIBN/ L; (○) [I]$_o$ = 0.01548 mol AIBN/L (9).

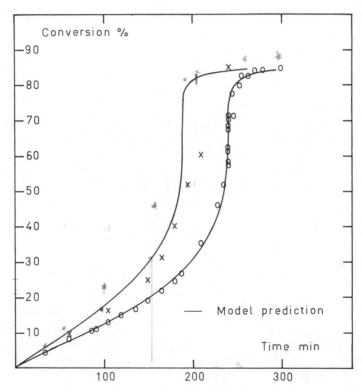

Figure 7.   Bulk polymerization of MMA at 50°C: ($\times$) $[I]_o = 0.05$ mol AIBN/L
(15); ($\bigcirc$) $[I]_o = 0.0258$ mol AIBN/L (9).

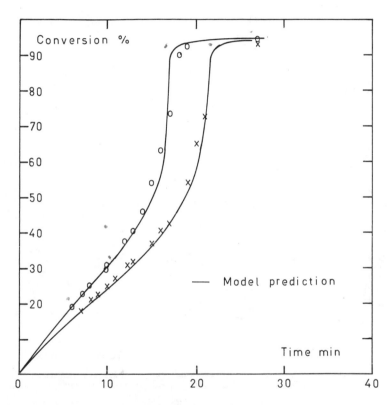

*Figure 8.   Bulk polymerization of MMA at 90°C: (○) $[I]_o = 0.0258$ mol AIBN/L;
(×) $[I]_o = 0.01548$ mol AIBN/L (9).*

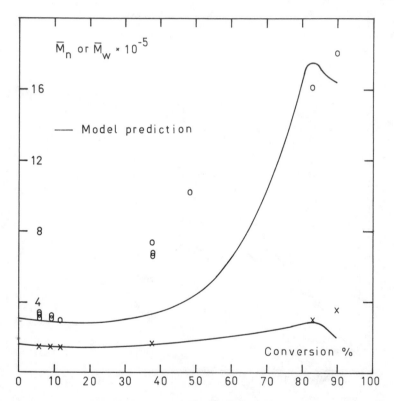

Figure 9.   Bulk polymerization of MMA at 70°C: effect of conversion on molecular weight averages. $(\times)$ $\overline{M}_n$; $(\bigcirc)$ $\overline{M}_w$. $[I]_o = 0.01548$ mol AIBN/L (9).

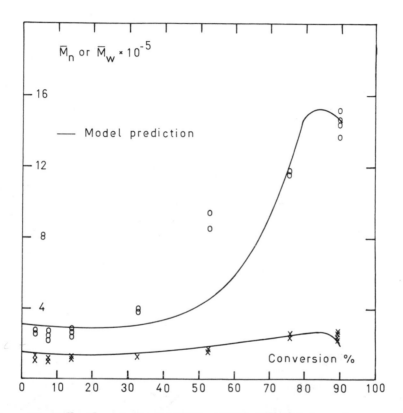

Figure 10.   Bulk polymerization of MMA at 70°C.   Effect of conversion on mo-lecular weight averages.   ($\times$) $\overline{M}_n$; ($\bigcirc$) $\overline{M}_w$. $[I]_o = 0.0258$ mol AIBN/L (9).

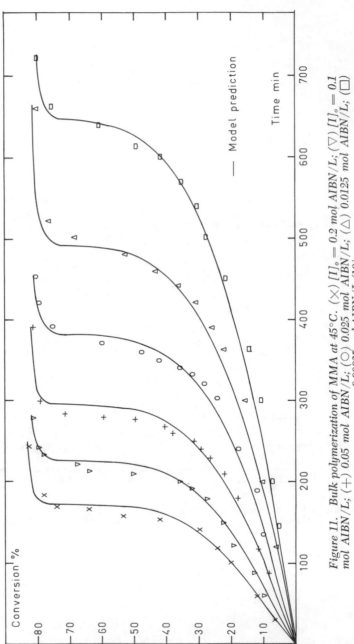

Figure 11. Bulk polymerization of MMA at 45°C. (×) $[I]_o = 0.2$ mol AIBN/L; (▽) $[I]_o = 0.1$ mol AIBN/L; (○) 0.05 mol AIBN/L; (+) 0.05 mol AIBN/L; (○) 0.025 mol AIBN/L; (△) 0.0125 mol AIBN/L; (□) 0.00625 mol AIBN/L (10).

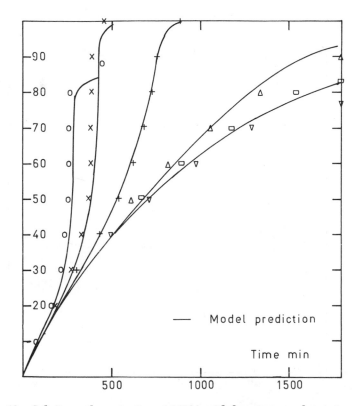

*Figure 12. Solution polymerization of MMA with benzene as solvent: tempera-
ture 50°C; $[I]_o = 0.0413$ mol BOP. ($\bigcirc$) zero, Benzene $= B$, 1.030 L MMA; ($\times$)
0.206 L B, 0.824 L MMA; ($+$) 0.412 L B, 0.618 L MMA; ($\triangle$) 0.618 L B, 0.412 L
MMA; ($\square$) 0.824 L B, 0.206 L MMA; (∇) 0.927 L B, 0.103 L MMA (16).*

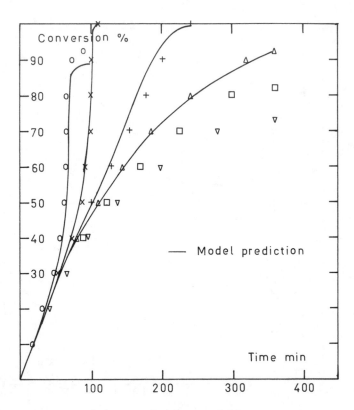

Figure 13.   Solution polymerization of MMA with benzene as solvent: temperature 70°C; [I]$_o$ = 0.0413 mol BOP. (○) zero Benzene = B, 1.057 L MMA; (×) 0.211 L B, 0.846 L MMA; (+) 0.423 L B, 0.634 L MMA; (△) 0.634 L B, 0.423L MMA; (□) 0.846 L B, 0.211 L MMA; (▽) 0.915 L B, 0.142 L MMA (16).

Nomenclature

A	constant
B	constant
C_M	chain transfer constant to monomer
C_S	chain transfer constant to solvent
D_M	diffusion coefficient of the monomer
$D_{M_{cr}}$	diffusion coefficient of the monomer at the conversion where the propagation becomes diffusion controlled
D_p	diffusion coefficient of a polymer radical
$D_{p_{cr}}$	diffusion coefficient of a polymer radical at the conversion where the termination becomes diffusion controlled
d_M	density of monomer
d_P	density of polymer
f	initiator efficiency
$[I]_o$	entanglement constant
K_3	temperature dependent constant
k_d	decomposition rate constant
k_{p_o}	propagation rate constant at zero conversion
k_p^o	propagation rate constant
k_{t_o}	termination rate constant in the absence of gel effect
k_t^o	termination rate constant
k_{t_c}	termination rate constant between entangled and non-entangled radical
$k_{t_{cr}}$	termination rate constant of the conversion where the termination becomes diffusion controlled
k_{t_e}	termination rate constant between two entangled radicals
k_1	temperature dependent constant
$\tilde{k}_1$	temperature dependent constant
k_2	constant
$\tilde{k}_2$	constant
k_4	constant
M	molecular weight of monodispersed polymer
[M]	monomer concentration
$\bar{M}_N$	number average molecular weight
M_o	molecular weight of monomer
M_w	weight average molecular weight

$\bar{M}_{wcr1}$ weight average molecular weight of the conversion where the gel effect starts

m   constant

n   constant

Q   constant

R   gas constant

[s]   concentration of solvent

T   polymerization temperature

T_{g_M} glass transition point of monomer

T_{g_P} glass transition point of polymer

T_{g_S} glass transition of solvent

$T_{g\infty}$ glass transition point of polymer with infinite molecular weight

t   time

V_F free volume fraction

$V_{F_{cr1}}$ free volume fraction at the conversion where the gel effect starts

$V_{F_{cr2}}$ free volume fraction at the conversion where the propagation becomes diffusion controlled

V_M volume of monomer

V_P volume of polymer

V_S volume of solvent

V_T total volume

$\bar{X}_N$ number average degree of polymerization

$\bar{X}_W$ weight average degree of polymerization

x   conversion of monomer

x_c number average chain length at the point of chain entanglement

Greek Symbols

α constant, exponent

α_g expansion coefficient for the glassy state

α_ℓ expansion coefficient for the liquid state

β constant, exponent

ε volumetric contraction coefficient

δ jump distance

δ_2 jump distance

Ψ_1 temperature dependent lumped constant

Ψ_2 temperature dependent lumped constant

Ψ_3 lumped constant

η viscosity

τ the reciprocal instantaneous number average degree of polymerization

ϕ_p volume fraction of polymer

ϕ_o jump frequency

ϕ_2 jump frequency

Literature Cited

1. Cardenas, J. and O'Driscoll, K.F., J. Polym. Sci. (1976) A-1, 14, 883.

2. Cardenas, J. and O'Driscoll, K. F., J. Polym. Sci. (1977) A-1, 15, 1883.

3. Cardenas, J. and O'Driscoll, K.F., J. Polym. Sci. (1977) A-1, 15, 2097.

4. Friis, N. and Hamielec, A.E., ACS Symposium Series (1976) 24, 82, "Gel Effect in Emulsion Polymerization of Vinyl Monomers".

5. Berens, A.R., ACS Symposium Series (1978) 39, 236, "The Sorption of Gases and Vapors in PVC Powders".

6. Hayden, P. and Melville, Sir Harry, J. Polym. Sci. (1960) 43, 201.

7. Beuche, F., Interscience, New York (1962), "Physical Properties of Polymers".

8. Abuin E. and Lissi, E.A., J. Macromol. Sci. Chem. (1977) A-11, 287.

9. Balke, S.T. and Hamielec, A.E., J. Appl. Polym. Sci. (1973) 17, 905.

10. Ito, K., J. Polym. Sci. (1975) A-1, 13, 401.

11. Horie, K., Mita, I. and Kambe, M., J. Polym. Sci. (1968) A-1, 6, 2663.

12. Balke, S.T., "The Free Radical Polymerization of Methyl Methacrylate to High Conversion", Ph.D. Thesis, McMaster University, Hamilton, Ontario (1972).

13. Abdel-Alim, A.H. and Hamielec, A.E., J. Appl. Polym. Sci. (1972) 16, 783.

14. Polymer Handbook (2nd ed.), Brandrup,J. and Immergut, E.H., Wiley, New York (1975).

15. Nishimura, N., J. Macromol. Sci. (1966) 1, 257.

16. Schulz, G.V. and Harborth, G., Makromol. Chem. (1947) 1, 106.

17. Duerksen, J.H., "Free Radical Polymerization of Styrene in
     Continuous Stirred Tank Reactors", Ph.D. Thesis, McMaster
     University, Hamilton, Ontario (1968).
18. Ibragimov, I.Y. Bort, D.N. and Efremova, V.N.,   Vysokomol.
     Soedin. Ser.B, (1974) 16 (5), 376.
19. Rudd, J.F., J. Polym. Sci. (1960), 44, 459.
20. Bueche, F., J. Polym. Sci., (1960), 43, 527.

RECEIVED February 9, 1979.

Technology of Styrenic Polymerization Reactors and Processes

R. H. M. SIMON and D. C. CHAPPELEAR

Monsanto Company, Springfield, MA 01151

1. Introduction

In considering the broad commercial applications of both crystal polystyrene (PS) and rubber modified "high impact" polystyrene (HIPS), it bears reemphasis that the process and process conditions each have major effects on product properties and fabrication behavior as well as product costs. With crystal polystyrene, product molecular weight, molecular weight distribution, oligomer and residual monomer levels, color and clarity are closely process related. With HIPS, rubber phase particle size, size distribution and morphology, graft copolymer level and molecular weight are additionally affected. To the manufacturer, therefore, the selection of the optimum process and conditions will underly the most relevant polymer "property": the cost of the product which meets performance requirements.

A characteristic of styrene polymerization processes is that different reactor types are frequently used in varying series combinations. The goal of this review is therefore twofold: first, to describe how and why different reactors have been employed in batch and continuous processes; and second, to outline some of the bridges between available theory and actual practice by highlighting some of the major design problems that are amenable to such an approach. Hopefully, this may encourage more pertinent research in the area. Industrial practice is reflected in the patent art, and a few general reviews such as Bishop (1). Much information remains proprietary. Answers to many practical problems have to be obtained by licensing or extensive development.

2. Processes and Reactor Process Elements

2.1 <u>Classification of Processes and Reactors.</u> Most styrene polymers are produced by batch suspension or continuous mass processes. Some are produced by batch mass processes. "Mass" in this sense includes bulk polymerization of the polymer dissolved in its monomer and, in some cases, some amount of solvent. PS mass polymerization is homogeneous (single phase viscous fluid). In mass HIPS polymerizations, the rubber forms an emulsified second phase.

Table I provides an overview of general reactor designs used with PS and HIPS processes on the basis of reactor function. The polymer concentrations characterizing the mass polymerizations are approximate; there could be some overlapping of agitator types with solids level beyond that shown in the table. Polymer concentration limits on HIPS will be lower because of increased viscosity. There are also additional applications. Tubular reactors, for example, in effect, often exist as the transfer lines between reactors and in external circulating loops associated with continuous reactors.

Various reactor combinations are used. For example, the product from a relatively low solids batch-mass reactor may be transferred to a suspension reactor (for HIPS), press (for PS), or unagitated batch tower (for PS) for finishing. In a similar fashion, the effluent from a continuous stirred tank reactor (CSTR) may be transferred to a tubular reactor or an unagitated or agitated tower for further polymerization before devolatilization.

Greater detail will be provided in the sections following.

TABLE I
Styrene Polymer Reactors - Classification

Reactor Function	Process Type	
	Batch	Continuous
Mass Polymerization	Conventional kettle with:	CSTR with: Agitated towers
Polymer <20% concentration	turbine agitator	turbine agitator Tubular reactors
Polymer 20-50% concentration	large turbine, anchor or helical agitator	turbine, anchor or helical agitator
Polymer 30-80% concentration	anchor or helical agitator proprietary and patented stirred reactors	anchor, helical agitators or special designs
Polymer >80% concentration	press, unagitated batch tower	Tubular reactors Unagitated towers
Suspension	Conventional kettle with turbine agitator	No commercial application

2.2 Batch Processes

2.2.1 Batch-Mass Reactors. The batch-mass reactors
used in these processes are of two types: low conversion agitated
kettles and high conversion static reactors with extended cooling
surfaces.

An example of a low conversion reactor would be a conventionally agitated kettle with large turbine agitators and jacket
cooling. The utility of this type of reactor can be extended to
intermediate conversions by the use of anchor or helical agitators
to partially overcome heat transfer and mixing problems at higher
viscosities.

A well-known high conversion reactor is the so-called
polymerization press, a modified plate-and-frame filter press
where polystyrene is polymerized in frames alternating between
cooling platens through which water (or steam) can be circulated.
Other versions of the high conversion reactor have been utilized,
e.g., the early "can process" of Dow, where styrene monomer was
placed in sealed cans in water baths and the metal stripped off
at the end of the polymerization (2).

Although low conversion reactors can be used for PS and
HIPS, high conversion batch reactors are generally limited to PS
because of difficulties with HIPS. In particular, the HIPS cake
from a polypress is difficult to grind and, because of poor
temperature control, is inferior in toughness.

Process flow for a typical batch-mass polystyrene process(1)
is shown in Figure 1. Styrene monomer is charged to the low conversion prepolymerization reactor with catalyst and other additives, and the temperature is increased stepwise until the desired
conversion is reached. It is then transferred into the press.
Polycycles are 6 to 14 hours in the low conversion reactor, and
16 to 24 hours in the press. At completion, the cakes are then
cooled with water and removed from the press to be ground and
then (usually) extruded into pellets.

The prepolymerization reactor for HIPS is similar (1). A
solution of rubber and styrene monomer is charged to the reactor
along with catalysts, antioxidants, and other additives, and the
temperature program is carried out until the desired conversion
is reached. This is usually close to the point where increasing
viscosity seriously limits mixing and temperature control.
Because of the difficulties of presses with HIPS cited earlier,
it is usual to transfer the syrup to a suspension reactor
containing water and a suspending agent for the completion of
polymerization. Design problems for suspension reactors will be
discussed in the next section. Design problems for HIPS prepoly
batch-mass reactors are analogous to HIPS continuous reactors as
discussed in Section 2.3.

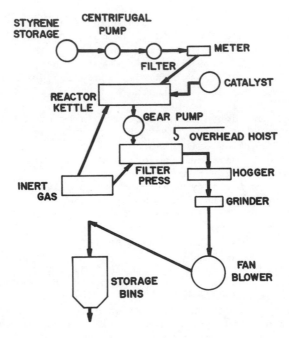

Figure 1.   Batch-mass process flow diagram (1)

Major reactor problems in batch-mass reactors are:

1) Heat removal/temperature control
2) Materials handling
3) Rubber particle size control (in HIPS)

Temperature control limits the rate of polymerization and the maximum conversion that can be obtained. Because of the high viscosities reached with polystyrene-monomer solutions, temperature control is frequently relatively poor, with temperature varying from place to place in the reactor (hot spots), and with time (peaking). This variation of temperature leads to a broadening of the molecular weight distribution (MWD) as discussed in the next section. If loss of cooling occurs, due to agitation failure or other causes, the reaction will run away. It is not normally practical to contain a runaway starting from these relatively low conversions. Adequate provision for venting of the high viscosity solution must be made.

In batch reactors, heat transfer will also limit the rate of heat-up to the required temperatures for initiation of polymerization. Use of a multiple catalyst system to provide lower temperature initiation has been proposed to minimize the time and energy required in heating.

Practical considerations of unloading play a major role in batch-mass reactors. For "low" conversion reactors, a suitably sized dump line must be chosen, based on acceptable N_2 dump pressure in the reactor, or discharge pump capability, as well as consideration for limiting reaction in the line between batches.

For a polypress reactor, control of cake sticking and provision for unloading the cakes are practical considerations which can outweigh any theoretical kinetic determinations.

Finally, for HIPS, control of the rubber morphology in its broadest sense requires control of the reactor environment as well as the chemistry of the grafting reaction. Poorly agitated regions, for instance, can lead to visual and physical defects in the product.

2.2.1.1 Peaking and Non-isothermal Polymerizations.

Biesenberger et al (3) have studied the theory of "thermal ignition" applied to chain addition polymerization and worked out computational and experimental cases for batch styrene polymerization with various catalysts. They define thermal ignition as the condition where the reaction temperature increases rapidly with time and the rate of increase in temperature also increases with time (concave upward curve). Their theory, computations, and experiments were for well stirred batch reactors with constant heat transfer coefficients. Their work is of interest for understanding the boundaries of stability for abnormal situations like catalyst mischarge or control malfunctions. In practice, however, the criterion for stability in low conversion

stirred reactors is that the potential for heat removal remains
sufficiently above the heat generation rate to give an adequate
margin of safety throughout the batch. The worst condition for
batch mass-suspension processes is normally at the maximum
conversion prior to transfer, where the heat transfer is lowest.
Fig. 2 shows temperature vs. time plots for polystyrene polymerization with two slightly different initiator concentrations (3).
One shows thermal ignition (and runaway reaction) while the
other has a mild temperature peak with less than 20°C overshoot.
The analysis is based on the superposition of a sigmoidal heat
generation curve as a function of temperature and the linear
heat removal curve as shown in Fig. 3. These curves intersect at
three equilibrium points: the bottom and top ones are stable, and
the center one is unstable and never utilized. The heat generation curve flattens out at very high temperatures due to the
reverse reaction, depolymerization. In practice, polymerization
reactors are operated on the lower portion of curve. If the heat
removal rate exceeds the heat generation rate in this region,
the reaction will trend towards the lower equilibrium point.

In practice, temperature control is used on the
coolant to maintain the reaction temperature at a higher,
commercially acceptable, level while maintaining control. When
the minimum coolant temperature is reached, however, any further
increase in reaction rate will lead to ignition.

In general, for polymerization reactions, the heat
generation rate is not a single-valued function of temperature,
g(t), but also a function of monomer and catalyst concentrations,
f(c). This is particularly important in high conversion
reactions where a certain amount of peaking can be tolerated.
The temperature excursion is limited because of depletion of
catalyst or, finally, by the depletion of monomer.

Depletion of catalyst (dead-ending) at constant or
increasing temperature has been modelled by Biesenberger (4).
Styrene polymerizations were modelled for a well-stirred batch
reactor with constant heat transfer and no thermal initiation.
Starting with monomer, the model predicted either quasi-iso-
thermal or quasi-adiabatic behavior depending on initial
initiator concentration (azo-bis-isobutyronitrile) and temperature. In the isothermal cases, the instantaneous degree of polymerization, X_n, varied, with monomer depletion tending to decrease
it and the gel effect increasing it. When dead-ending occurred,
this also increased X_n. The greater the variation, the broader
the MWD (See Figs. 4, 5 and 6). In adiabatic and quasi-
adiabatic cases, dead-ending always occurred. The instantaneous
X_n first fell with increasing temperature; X_n then increased
rapidly as the initiator was depleted (dead-ending). MWD was
much broader in the adiabatic cases.

Although studied for agitated reactors, the
phenomena of thermal ignition are probably of more interest in
the nonagitated high conversion reactors such as the polymeriza-

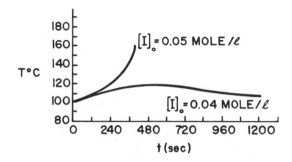

Figure 2. Experimental temperature–time profiles for batch styrene polymerization with and without thermal ignition (3)

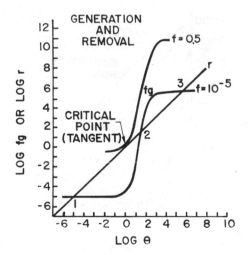

Figure 3. Dimensionless heat generation function fg and removal function r vs. dimensionless temperature θ at different values of a time-decaying parameter f (3)

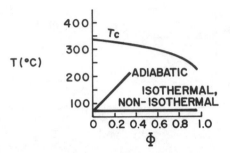

Figure 4.   Temperature vs. conversion for batch-mass styrene polymerization (4)

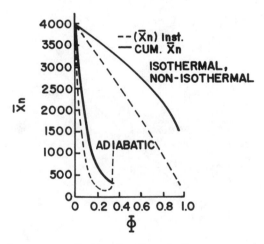

Figure 5.   Number average degree of polymerization vs. conversion for batch-mass styrene polymerization (4)

tion press. The heat generation rate forms a looping trajectory
when plotted against temperature as the reaction rate first
rises and then falls as monomer or catalyst is depleted, finally
approaching the heat removal curve at the final low polymerization
rate. See Fig. 7 (3). To model actual styrene polymerization
behavior at high temperature, the effects of chain transfer to
monomer and thermal initiation should be included, as well as
variations of heat transfer coefficient with temperature and
conversion.

2.2.1.2 Low Conversion Reactors. The major
problem in temperature control in low conversion reactors is the
orders of magnitude increase in viscosity as the conversion
increases. Fig. 8 shows the viscosity of a polystyrene
solution as the function of percent PS (5). The data are for
polystyrene with a Staudinger molecular weight of 60,000 at $100^{\circ}C$
and $150^{\circ}C$ in a cumene solution, a satisfactory analog for styrene
monomer solutions. As the polymer concentration increases from 0
to 60%, viscosity increases from about 1 cp to 10^5 cp.

In designing an agitated low conversion reactor,
the heat transfer and polyrate must be balanced at the worst
condition, normally the highest conversion expected, since heat
transfer decreases with increasing viscosity. A suitable safety
factor should be allowed for batches with abnormally high polyrates and delays in batch transfer.

Conventional turbine agitators are not normally
satisfactory at the viscosities involved. Metzner (6) found
that in high viscosity Newtonian and non-Newtonian solutions,
a flat-bladed turbine moves material only in the vicinity of
the impeller, and that the flow pattern was streamline. The
material near the wall was stationary. The use of relatively
large diameter turbines or paddles can extend their effectiveness
to higher viscosities. The heat transfer to a jacket in the
vessel agitated by a turbine in the turbulent region is given
by the following equation (7):

$$\frac{h_j D_j}{k} = .36 \left(\frac{L^2 N \rho}{\eta}\right)^{2/3} \left(\frac{c \eta}{k}\right)^{1/3} \left(\frac{\eta}{\eta_w}\right)^{.14}$$

This indicates that the heat transfer coefficient h_j
varies inversely with the 1/3 power of the viscosity η. Applying
the previously mentioned viscosity correlation, one can
determine that the heat transfer coefficient will decrease by
40-50% with every 10% increase in polystyrene conversion
between 0 and 40%.

Ide and White (8) studied the viscoelastic effects
in agitating polystyrene solutions with a turbine. At
concentrations below 30% PS, flow was normal. Above
35%, the viscoelastic forces caused the flow to reverse, moving
away from the impeller along the axis. At 30 to 35% PS, both
occurred, causing a segregated secondary flow around the turbine.

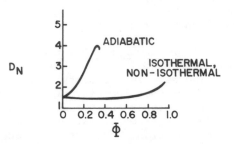

Figure 6.   Cumulative dispersion index vs. conversion with batch-mass styrene polymerization (4)

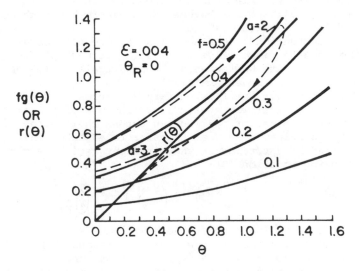

Figure 7.   Computed curves of dimensionless heat generation fg (θ) and heat removal r (θ) functions showing stable behavior of dimensionless temperature θ for nonautonomous systems (3)

Because of their proximity to the wall, anchor
agitators are more effective in maintaining heat transfer coefficients to higher viscosities. Relatively high torques and
larger gear reducers are required, however. Studies by Uhl
and Voznick (9) correlated the heat transfer coefficient in a
manner similar to that used with turbine agitators:

$$\frac{h_j D_j}{k} \propto \left(\frac{D_1^2 N \rho}{\eta}\right)^n \left(\frac{c\eta}{k}\right)^{1/3} \left(\frac{\eta}{\eta_w}\right)^{.18}$$

At Reynolds numbers above 400, the exponent n on
the Reynolds number is .61, very close to the .67 in the case of
a turbine. At Reynolds numbers below 400, where the flow is
presumably laminar, the exponent drops to .43. As a result,
heat transfer coefficient varies inversely with the viscosity
to the .28 and .1 power at Reynolds numbers above and below 400
respectively.

Heat transfer can, of course, be increased by
increasing the agitator speed. An increase in speed by 10^3 will
increase the relative heat transfer by 10^2. The relative power
input, however, will increase by 10^7. In viscous systems,
therefore, one rapidly reaches the speed of maximum net heat
removal beyond which the power input into the batch increases
faster than the rate of heat removal out of the batch. In
polymerization systems, the practical optimum will be significantly below this speed. The relative decrease in heat transfer
coefficient for anchor and turbine agitated systems is shown
in Fig. 9 as a function of conversion in polystyrene; this was
calculated from the previous viscosity relationships. Note that
the relative heat transfer coefficient falls off less rapidly
with the anchor than with the turbine. The relative heat
transfer coefficient falls off very little for the anchor at low
Reynolds numbers; however, this means a relatively small
decrease in an already low heat transfer coefficient in the
laminar region. In the regions where a turbine is effective,
Uhl and Voznick (9) found that the anchor and the turbine gave
roughly the same overall heat transfer coefficient at equal power
inputs. Turbines, however, are conventionally run with lower
power inputs than anchors. In the design of an anchor agitator,
it is interesting to note that for Newtonian fluids, the heat
transfer coefficient goes through a minimum at a clearance to
diameter ratio of approximately .02. With wider clearances, not
only is the heat transfer improved at a given agitator speed
but the power required is reduced (9). Mixing with an anchor
agitator becomes ineffective with viscosities over 10^5 cp.

Helical agitators have been developed to provide
good turnover and satisfactory heat transfer at viscosities
as high as 5×10^5 cp.

Coyle et al (10) showed that the heat transfer
coefficient for helical agitators remained in the region of 3-5

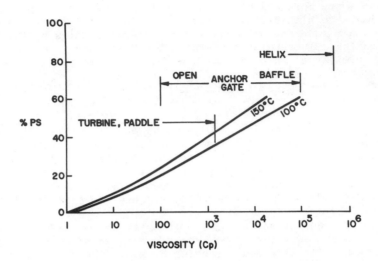

Figure 8.   Viscosity of polystyrene solutions ($M_{st} = 60,000$) (5) and recommended
ranges of different agitators (9, 10)

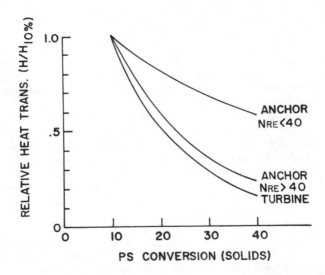

Figure 9.   Relative heat transfer coefficient vs. percent polystyrene solids based
on viscosity relationship shown in Figure 8

BTU/hr. sq.ft.$^\circ$F over a wide range of viscosities and rotational
speeds. This is equivalent to the thermal resistance of a fluid
film equal to about 1/2 the clearance between the helical agitator
and the vessel wall. This represents Reynolds numbers in the
range of 10^{-1} to 10. This is the region of creeping flow where,
with no inertial effects, there is little displacement of the
fluid adjacent to the wall.

The helical agitator has the added advantage of
good top-to-bottom mixing compared to the anchor. In one
experiment, this reduced the maximum temperature variation in the
batch from 10-20°F with an anchor to 5-15°F with a helix at a
viscosity of 4000 cps. With an anchor agitator, some additional
means of providing satisfactory top-to-bottom mixing may be
required at the higher viscosities. Uhl and Voznick (9)
indicate that the effectiveness of the anchor can be increased
by the use of a "gate" of cross members across the anchor and
even more by the use of a baffle inside the anchor.

Sawinsky et al (11) have correlated power requirements for Newtonian and psuedoplastic fluids agitated by
anchor and helical agitators.

A final consideration in addition to heat transfer
and blending is the flow regime within the vessel. This may be
important particularly for rubber modified polystyrene in pre-
venting the formation of macrogel and the control of particle
size. The flow intensity is normally characterized in terms
of the apparent shear rate although extensional (hyperbolic)
flow may be at least as an important as shear flow in particle
breakup. One can characterize an agitation system by how high
the maximum shear is and how often a given element of fluid
would pass through the high shear zone. With a conventional
turbine agitator, the shear rate will be approximately proportional to the tip speed; this will vary with ND_L, the agitator
speed times turbine diameter.

The time required to pass through the high shear
zone will be proportional to the volume of the vessel divided
by the flow from the turbine. This will be porportional to
$1/ND_L^3$. With a turbine in turbulent flow, turnover is relatively
rapid and all the fluid will pass through the impeller region
in a relatively short period of time. The flow regime in an
anchor or helically agitated vessel can be inferred from the
flow studies by Smith and Peters (12, 13, 14). These indicated
nearly creeping flow with no obvious inertial effects at
Reynolds numbers below 10-20. Streamlines at a Reynold number
of 2 are symmetrical around the blades, indicating reversible
flow and thus no pressure buildup in front of the blade.   From
this one can calculate the flow under the blade from the wedge-
shaped velocity profile arising from pure drag flow.   The
thickness of the relatively stagnant layer which passes between
the blade and the wall will increase as the blade approaches
and decrease again to the thickness equal to half of the

clearance (Fig.10). Such low Reynolds numbers are probably
unusual with anchor agitators but more common with helical
agitators. This would explain the observed independence of heat
transfer coefficient from Reynolds number with helical agitators
and is in agreement with the observed thermal resistance of a
layer half as thick as the clearance. As the Reynolds number
increases to 10, the flow pattern is generally the same, but no
longer quite symmetrical around the blade, indicating some
inertial (pressure) forces.

At Reynolds number above 10-20, a new flow regime is
established. While still laminar, it is obviously unsymmetrical.
The inertial effects, therefore, are important but do not
dominate to the extent of making the flow turbulent. In this
region, a stable trailing vortex is set up behind each blade.

At Reynolds numbers significantly greater than 10,
pressure flow due to the higher pressure in front of the agitator
will be superimposed on the drag flow between the agitator and
the wall (Fig. 11). The thickness of the layer, after passing
by the agitator, increases with increasing Reynolds number.
The streamlines of this flow are shown by Peters and Smith (12).
In this case, the effective thickness of this layer appears
to be about equal to the gap with the wall, indicating a
pressure flow about equal to the drag flow. It can be calculated
that this would increase the maximum shear rate on the fluid
passing under the agitator blade by a factor of seven.

The region of stable trailing vortices was sustained
to Reynolds numbers up to 5000 (14). Other flow regimes were
observed at higher Reynolds numbers but are probably not
achieved in polystyrene reactors.

The revolutions required for a volume of liquid
equal to the vessel volume to pass under the agitator can be
calculated from a knowledge of the geometry and the thickness
of the layer passing under the agitator. This number would be
misleading, however, since the fluid in this wall layer is
displaced only slowly by the secondary (top-to-bottom) flow.
Forty revolutions of a helical agitator are required to blend
liquid. A higher number of revolutions are required for blending
with an anchor agitator because of poor top-to-bottom mixing.
Peters (53) found 55 to 60 revolutions required for mixing with
an anchor at low Reynolds numbers, and a clearance to diameter
ratio of .04. When the clearance was decreased to .01, about
25% more revolutions were required. At higher rpm, mixing time
as well as revolutions increased. The system used was poly-
acrylamide. Viscoelastic behavior appeared to help mixing.

2.2.1.3 High Conversion Batch-Mass Reactors. Because
of the very high viscosities at high conversion, these reactors
are unagitated. Temperature control therefore depends upon
conduction through the polymer to extended heat transfer surfaces.
Most common are the cooled plates of the plate and frame

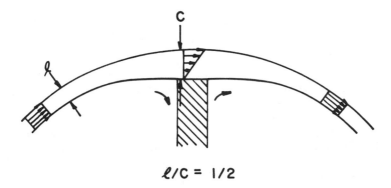

$$\ell/C = 1/2$$

Figure 10.   Pure drag flow of polymer syrup in the wall-blade clearance C of an anchor agitator in creeping flow.   All velocities relative to the blade (12).

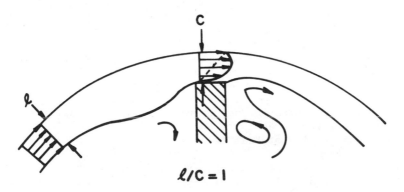

$$\ell/C = 1$$

Figure 11.   Flow patterns with an anchor agitator as in Figure 10 but at higher Reynolds numbers (above 10–20) where inertial effects become significant. Streamline schematic (12) shows stable trailing vortex.

polymerization press. Tube bundles, cans, etc., have also been used. The high heat of polymerization of styrene (17.6 kcal./gm. mole at 127°C) in combination with low thermal conductivity of polystyrene (about 2.9 cal./cm°C. at 80°C) leads to non-uniform temperatures in commercial reactors. Excessive distance between heat transfer surfaces must be avoided and long polycycles must be used to prevent runaway reactions and excessive hot spots in the regions at the greatest distance from the heat transfer surfaces. This problem is so severe that non-uniform polymerization temperatures may even arise in glass tubes used for laboratory kinetic studies if the tube diameter is not sufficiently small.

Bishop (1) has described the polycycle for styrene polymerization in the press. Thermocouples in different frames show the initiation of peaks in different chambers at different times. When the last cake has started to peak, full cooling water is applied to the press. Despite this cooling, with the relatively poor heat transfer to the polymer, the temperature in the center of the cake will continue to rise to 150 to 160°C. As the reaction rate falls off, steam is put on the press to raise the polymer to a higher temperature and bring the reaction close to completion.

The effect of this peaking in broadening MWD will be similar to ignition in stirred reactors previously described (3). Starting with prepolymerized syrups, the reaction becomes monomer limited sooner. Temperature gradients through the cake add another dimension to the calculation of MWD. The greatest temperature gradients will occur during the peak period when the polymer at the wall will be close to the water temperature, and the center of cake 150°C or more.

Partial differential equations can be solved showing the theoretical temperature distribution as a function of time and distance for various geometries such as parallel plates of a press polymerization, arrays of heat exchanger cooling tubes, etc. (15). With the large temperature gradients realized, however, natural convection may occur. Theoretical calculations ignoring this may underestimate the overall heat transfer. In practice, the spacing of heat transfer surfaces has been arrived at empirically.

Depletion of the catalyst during peaking or finishing would be predicted to lead to dead-ending (4). Thermal polymerization will be effective in increasing the conversion to some degree. Since thermal initiation of styrene is second-or third-order with monomer, the rate of polymerization falls off very rapidly as the monomer concentration falls below 10%. High temperature initiators and other proprietary or patented additives are used to drive the polymerization closer to completion. Higher residual monomers are still characteristic of high conversion batch-mass polystyrene when subsequent devolatilization is not used.

2.2.1.4 Access to Practice. Publications and patents on the batch mass process are limited. Bishop's book (1) contains the most detailed description of the polymerization press and mass-suspension processes for PS and HIPS. Fong (16) presents an economic analysis of the press process based on Bishop's description. Patent references are few for the batch-mass process: the 1939 Bakelite patent on transfer of prepoly syrup to chambers or containers is of historical interest (17).

Both Bishop (1) and Fong (16) give extensive patent reviews of the mass-suspension process for HIPS including the pioneering patents of Stein and Walter (18).

2.2.2 Suspension Polymerization. Suspension polymerization probably remains the most widely practiced method of producing PS. It can also be used to produce HIPS. To improve quality of the latter, however, a batch-mass pre-polymerization of the rubber syrup is normally carried out first; the syrup is then suspension polymerized to completion.

The advantage of suspension processes over mass processes is the excellent temperature control that can be obtained through the suspending medium, water. This allows for rapid heat removal and shorter polymerization times. It reduces or eliminates hot spots or heat-kicks characteristic of mass reactors. It also allows the polymerization to be driven very close to completion so that no devolatilization step is normally required.

There are disadvantages, however, to using water. It must be suitably pure, requiring deionization in many cases. The water must be heated and cooled, and the effluent containing suspending agents and a small amount of polymer emulsion must be treated before discharge in an increasing number of cases. Cost of suspending agents are additional factors. Suspending agents also reduce the clarity of the product compared to those from mass reactions.

The approximately round shape and small size of the suspension beads is useful for some applications such as expandable polystyrene or as an intermediate for further compounding with pigments, other polystyrene beads, etc. Being round, however, they tend to roll, not only causing a safety hazard when spilled on floors but more importantly causing difficulties in some fabricating extruders and molding machines. Except for expandable polystyrene, beads are seldom sold as such but are extruded into pellets.

2.2.2.1 Process Description. A simplified process flow for a typical PS suspension process is shown in Figure 12. Storage for monomer, bead slurry and dried beads are not shown. Suitable storage, blending, extrusion and packaging equipment will also be required to produce finished pellets for sale. The suspension reactor is typically a 4,000-gallon glass-lined,

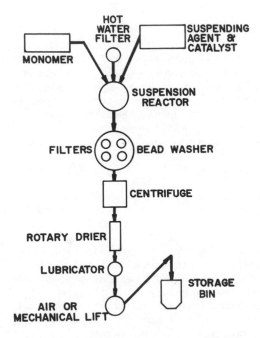

Figure 12.   Suspension polymerization flow diagram (1)

jacketed, agitated kettle. Larger, stainless steel clad reactors are increasingly common, as discussed in Section 2.2.2.3.

Styrene is metered to the reactor containing hot water. The batch is heated further through a controlled temperature cycle. Although chain transfer agents, such as mercaptans, can be employed for molecular weight control, this is not common with PS. Generally, molecular weight is controlled via the temperature cycle within the constraints of reactor heat removal capability. Batch temperature is generally raised near the end of the polymerization cycle to drive the reaction closer to completion. Suspending agent may be added initially or after the early phase of polymerization, depending upon the system. Other additives, such as initiators, plasticizers, tints, etc., may be charged to the reactor initially or at one or more points during the batch cycle. After completion of the polymerization, the batch is cooled well below the glass temperature of the beads (ca. 90-100°C) and dumped to a hold tank from which it is centrifuged to separate most of the suspension liquor. The beads are then dried, typically in a co-current hot air rotary dryer.

The process for rubber-modified polystyrene is similar, except that a prepolymerization syrup from a batch-mass reactor is the starting point.

2.2.2.2 <u>Major Suspension Reactor Problems</u>. The critical problems in defining or optimizing a suspension reactor can be divided into three areas. In approximate order of importance, these are:
1. Suspension stability
   a. Set-up
   b. Buildup
   c. Particle size
   d. Particle size distribution
2. Heat removal
3. Kinetic
   a. Molecular weight
   b. Residual monomer
   c. Cross-linking of rubber

There are obvious interactions between the first two areas. A set-up batch, if not resuspended or dumped, will lose heat transfer and probably run away nearly adiabatically. Not only should the agitation system and suspending agent be jointly designed to minimize this, but emergency suspending agent, dumping, and kettle venting systems should be designed for the worst case. Shortstop systems using sulfur or other free radical inhibitors are sometimes less effective in suspension system because of the difficulty of mixing them into the viscous beads where initiation and propagation take place. Buildup can occur either rapidly during an unstable batch or slowly over many normal batches. Buildup drastically reduces jacket heat transfer, slowing heatup and cooldown and, if serious

enough, leads to loss of batch temperature control. These
"heat kicks" will be self-limiting if monomer or catalyst is
depleted but can also lead to runaway conditions and emergency
venting. Conversely, excessive boilup due to venting,
stripping, or reflux cooling can lead to buildup above the
liquid level and even lifting of the batch and set-up. Bead size,
bead size distribution, and gaseous occlusions must be kept in
reasonable bounds for downstream processing (washing, drying, and
extrusion). In such cases as expandable polystyrene and ion
exchange resins, close control of particle size and distribution
are critically important. Even if downstream screening is
included, yields and productivity suffer from variability and
broad distributions.

The most critical art in suspension polymerization
concerns the suspending agent, as reflected by the patent and
scientific literature. A large number of protective colloids of
carefully controlled solubility and molecular weight as well as
inorganic solids of carefully controlled particle size and
surface wetting characteristics are used. Even if the suspending
agent is fully developed on the laboratory scale, problems will
often remain with scale-up, optimization, and/or debottlenecking.
The reactor agitation interacts so strongly with the suspending
agent that the two should be correctly thought of as one system.
Process variables such as phase ratio, temperature program, and
add schedule for monomers and other additives also have effects
on suspension stability.

2.2.2.3 Heat Transfer and Suspension Stability. The
second major reactor problem in suspension polymerization is
heat removal. Simon and Alford (19) have shown that jacket heat
removal is so closely coupled to suspension stability that
careful monitoring of the jacket heat transfer coefficient can
serve as an early warning to suspension failure. The traditional
indication of suspension failure is abnormally high agitator
drive power. However, since suspension reactors are normally
turbulent, major changes of viscosity can take place without
significant power changes. By the time buildup on or near the
agitator is sufficient to be detected in power, coalescence
may have progressed to the point where the chance of saving the
batch is reduced. Early warning from careful monitoring of heat
transfer coefficient can lead to early and more successful
corrective action. It is also important to know how the heat
transfer changes during the batch so allowance can be made for
this in programming the polymerization. This also gives some
insight into the mechanism of suspension. Apparently, as drops
begin coalescing with each other, they also coalesce with heat
transfer surfaces, significantly decreasing the heat transfer
coefficient.

Figure 13 shows the variation of heat transfer
coefficient as it falls off with increasing conversion and

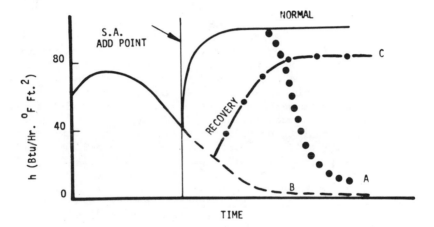

Figure 13.   Response of early warning probe heat transfer coefficient to different events during a suspension polymerization batch (19)

recovers when suspending agent is added. This coefficient is
calculated from Δ T across a cooling baffle with a constant flow
of cooling water running through it. The batch can fail to
suspend as shown by curve B, or fail after suspension as indicated by curve A. Emergency suspending agent addition can lead to
recovery or the batch as shown in curve C.

Heat transfer problems become more severe as reaction
rates are increased and water-to-monomer ratios are reduced. In
addition, as reactor sizes are increased for improved process
economics, the amount of wall heat transfer surface area per unit
volume will drop and result in a lower reactor space-time yield.

Jacketed reactor heat transfer limitations can be
avoided to varying degrees by several techniques. As outlined by
Beckmann (20), these include scraped surface heat exchange
to improve the coefficients, external cooling loops, cooling with
water or monomer injection, or reflux cooling. Thus, the once
standard reactors of 4,000 gal. nominal volume are givi g way to
reactors of 13, 26 and 54,000 gallons.

Temperature control can present special problems in
large reactors if the turnover is too low. This is caused in
part by the great hydrostatic difference, superheating the
material at the bottom. Beckmann (20) reports temperature
differences of as much as 10 to 20° in large reactors.

Chemische Werke Hüls has overcome some of the practical problems of large scale reactor design (20). Bottom entering
agitators and side-mounted baffles were used to overcome stress
problems in heavy agitator shafts. As indicated earlier, scale-
up at equal power per unit volume will greatly increase Reynolds
numbers and therefore give more turbulent behavior. Gravity
forces also become greater, decreasing the size of the vortex.
Both factors help in scale-up. Special high strength steels
were used to eliminate stress relieving after welding and to
minimize freight in shipment. With relatively thin shells at
large scale, stiffness of the reactor depends upon the jacket
design as well as the stainless clad reactor shell. Since
glassing of these large reactors is impractical, the stainless
steel cladding is electropolished to give a smoother surface to
prevent buildup. Also discussed are emergency shutdown by the
addition of the shortstop and emergency venting of vapors for
cooling. To this could be added emergency suspending agent
mentioned earlier. The venting described here is a controlled
venting of vapors to provide agitation and reflux cooling.
This must be done with care to avoid lifting the batch and
plugging the vent. To safeguard the kettle, a proper vent line
on the rupture disc must be provided. This must be sized to
allow relieving under the worst conditions of exothermic reaction
where a large volume of water vapor must be vented, as well as a
viscous liquid layer caused by loss of suspension. Fortunately,
the venting problem here is still not as severe as in mass
reactors.

Albright et al (21) reviewed the practical aspects of designing reactors, with particular attention to polystyrene and polyvinyl chloride. Overall heat transfer coefficients of 40-60 for glass and 50-110 BTU/hr.ft.^{2} °F for stainless steel are given. Any wall (or jacket) buildup will reduce this coefficient significantly. Cleaning manually, with high pressure water, or with solvents is often required even with glass or highly polished stainless steel reactor walls.

2.3 Continuous Processes

2.3.1 General Considerations. Continuous mass processes for polystyrene have been in commercial use since the 1930's, and for rubber-modified polystyrene (HIPS) since the 1950's. Much of the information on equipment design, process configuration and operating parameters connected with continuous processes is found in the patent literature. There are inherent limitations to such sources. Recognizing this, we will provide a survey of this subject here, concentrating on reactor selection and design considerations.

2.3.2 Reactor Types. Continuous reactors for PS and HIPS processes exist in widely varying designs, as shall be discussed in ensuing sections. However, they fall into two general classifications: the continuous stirred tank reactor (CSTR) and the linear flow-reactor (LFR). Each type has operating characteristics which may make it more or less appropriate for a given process function, and these must be most carefully considered by the process designer.

A CSTR is a deliberately backmixed reactor and, in principle, its effluent temperature and composition are the same as the reactor contents. With an ideal CSTR, the feed blends instantaneously with the uniform reactor contents. In actual practice, of course, we find that feed blending time may be protracted, and varying degrees of segregation, short circuiting and stagnation exist in the reactor contents.

The ideal LFR is a plug flow reactor, but true plug flow is not attained in commercial reactor designs due to velocity profiles and varying amounts of backmixing. However, in designing an LFR, the intent is to provide for a net, progressive change in composition and perhaps temperature as the reactants move through the reactor. Vertical LFR's are sometimes called towers.

CSTR's may be more appropriate during the earlier polymerization stages where viscosities are relatively low. Although designs have been developed which provide high degrees of backmixing even with polymer syrups with relatively high viscosities, practical limitations are ultimately reached, as discussed in Section 2.2.1.2. In addition, at the higher conversions corresponding to these elevated viscosities, reaction rates are usually decreasing. A single CSTR operating at such conversions might therefore be uneconomic due to low space-time yields. Multiple or staged CSTR's reduce this problem.

Since a CSTR operates at or close to uniform conditions of temperature and composition, its kinetic and product parameters can usually be predicted more accurately and controlled with greater ease. The CSTR can often be operated at a selected conversion level to optimize space-time yield, or where a particular product parameter is especially favored.

Excessive backmixing can be very difficult to prevent at low reactant viscosities, so an LFR may not be practicable where dissolved polymer solids are low, either due to low conversions or high solvent levels. Under these conditions, LFR behavior can be approached by incorporating a sufficient number of CSTR's in series.

As viscosity increases, backmixing in an LFR can be more easily reduced since flow conditions become more laminar and stratification can be more closely approached. Where the reactor is cooled via heat transfer surfaces, large areas must be provided for temperature control because heat transfer co-efficients will be low with the generally limited agitation levels. With a CSTR at the lower range of syrup viscosity, on the other hand, there are fewer limits on agitation intensity, especially in the PS processes. Higher heat transfer coefficients are achievable and it is therefore easier to design for higher heat fluxes than would be the case with an LFR.

LFR's are especially important where high conversions are to be achieved. Since the upstream zones of the reactor will be at lower conversions and, therefore operating at generally higher reaction rates, the space-time yield will be higher than a CSTR operating at the LFR exit conversion. Since it is in principle possible to divide an LFR into a number of axial temperature zones, and since additives can be charged at different axial points, the process designer will usually enjoy greater flexibility with product parameters than can be achieved with a CSTR. If the reactor operating parameters for two polymer products can be made to differ primarily in the downstream zones of an LFR, it is possible in principle to have less transition material when going from one product to the other than if a single CSTR were used.

Since LFR's generally run under temperatures, viscosities and other parameters varying as one proceeds downstream, it follows that process and product control can be more complicated than with a CSTR.

With the foregoing considerations in mind, we will now examine commercial CSTR and LFR designs and how they are incorporated in various commercial PS and HIPS processes.

2.3.3. I.G. Farben Process. The first continuous mass polystyrene process was developed in Germany by I.G.

Farbenindustrie A.G. during the 1930's (22) and was reported
on in detail shortly after World War II (23). It was developed
directly from a batch process.

A schematic drawing is shown in Fig. 14. Two "prepoly"
CSTR's in parallel, each with a 1400-kg. holdup, fed a total of
44 kg./hr. of syrup at 80°C and 33-35% conversion to a second-stage
LFR. The CSTR's were cooled via jackets and internal cooling
coils, and slowly agitated with gate-type agitators.

The unagitated tower LFR was divided into six
temperature controlled sections, each 1 meter long and equipped
with a jacket and a helical cooling coil. The temperatures
varied longitudinally from 100-110°C to 180°C. The effluent
melt was extruded into a sheet onto a stainless steel cooling
belt and then granulated. No catalyst was used but .02% glacial
acetic acid was added, reportedly to help reduce final residual
volatiles to .5%.

The reactors had relatively limited heat transfer
capability and polyrates therefore had to be kept low. This was
accomplished by operating at low temperatures. (The rate in the
CSTR's was about .5%/hr.). Since chain transfer agents were not
employed, product Staudinger molecular weight was about 100,000,
very high by current commercial standards.

The use of the CSTR and LFR by this process follows
the guidelines discussed in Section 2.3.2. The former is used
for the first polymerization stage where viscosity is relatively
low. The latter where viscosities are high enough to suppress
backmixing and where very high exit conversions are desired.
It is likely that some devolatilization occurred during the
extrusion step, possibly aided by the acetic acid, since it is
doubtful that there could be sufficient residence time in the
final 180°C tower section to drive the conversion to 99.5%.

2.3.4 Union Carbide Patents. During World War II,
Union Carbide and Carbon Corporation patented a process (24)
having some features in common with the German process described
above. The Union Carbide process as patented is illustrated in
Figure 15. The continuous process consists essentially of two
reaction zones followed by a milling-pumping zone.

The first reaction zone consists of a CSTR operating
at about 125°C with syrup exiting at 70 - 80% solids. The
complex, ruggedly built agitator is designed to provide constant
syrup circulation between the walls and shaft and to provide
some discharge pumping action. Although a jacket is provided
at the lower portion of this reactor, temperature control is
accomplished primarily by reflux cooling, with the reactor
being maintained under 18-22 inches Hg vacuum.

The patent suggests that the first reaction zone can
be replicated by three CSTR's in series with the first operating
at 100°C and about 35% solids, the second at 115-120°C and 65%
solids, and the third at 140°C and about 85% solids. This

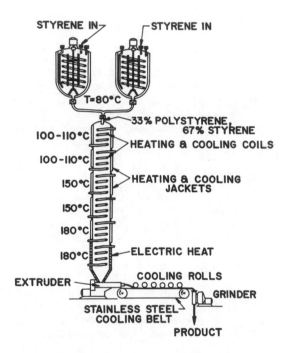

Figure 14.   I. G. Farbenindustrie continuous mass polystyrene process (23)

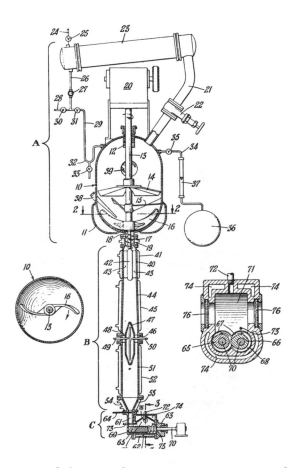

Figure 15.   Union Carbide patented continuous mass process for polystyrene (24)

configuration appears more desirable since the space-time yield
is probably improved significantly over the single reactor case
and greater flexibility is also possible. It has been estimated
(24) that styrene polyrates in the three CSTR's are respectively
about 2, 4, and 7.5%/hr., resulting in polymer production respectively of about 1.1, 2.2 and 4.2 lbs./ hr. per ft.^{3} of reaction
mixture. On this basis, one can estimate that the reactant
volumes are respectively in the ratio of about 6.7:2.9:1. The
higher surface-to-volume ratios of the smaller reactors can be
expected to ease heat transfer problems in face of increased
polyrates and reactant syrup viscosities. Agitator designs in
the three CSTR's of this first reaction zone are believed to be
quite different (25). The effluent from the first reaction
zone is pumped to the second zone consisting of a tower reactor.
The pumping action is accomplished by the first zone agitator
and/or a screw conveyor just downstream of the first reaction
zone.

    In the tower, the inlet syrup temperature is rapidly
raised to about 175°C with a heat exchanger. As the syrup
progresses down the tower, the heat of polymerization gradually
raises its temperature to 210-215°C by the time it discharges
at 95-97% solids.

    The syrup discharging from the tower enters an enclo-
sure where it is subjected to milling under vacuum and/or an
inert atmosphere. Vacuum as high as 29.8 in. Hg is suggested.
The action in this chamber accomplishes three purposes: a)
blends the polymer melt, b) removes most of the unreacted
volatile materials in the melt, and c) degrades the higher
molecular weight polymer fraction ($M_{st} \gtrsim 150,000$) without
substantial increase in the low molecular weight fraction
($M_{st} < 30,000$). The melt after milling is pumped through a
strand die, cooled and pelletized.

    As can be seen, the principal differences of this
process from the earlier German process includes: a) reflux
cooling of the first reaction zone, b) the possible division of
the first reaction zone into a series of CSTR's and c) the
incorporation of a devolatilization step.

    2.3.5 <u>Dow Process Developments</u>. The Dow technology
is taught in a large number of U.S. patents. Those by McDonald
<u>et al</u> (26), Ruffing <u>et al</u> (27), and Finch <u>et al</u> (28) are good
illustrations of this series of developments as is the summary
by Amos (2).

    The continuous polystyrene process which was commer-
cialized successfully in 1952 (2) is illustrated schematically
in Fig. 16.   It is characterized by three vertical elongated
reactors in series, the contents of which are gently agitated
by slowly revolving rods mounted on an axial shaft. Temperature
control is provided by horizontal banks of cooling tubes between
adjacent agitator rods. Such a reactor, called a "stratifier-

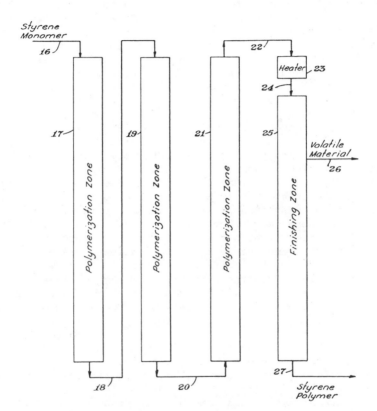

Figure 16.   Schematic of a Dow Chemical Co. patented continuous mass process for polystyrene (26)

polymerizer" (2, 26) is shown in Fig. 17. The revolving rods
prevent channeling, promote plug flow, and aid in heat transfer.
With HIPS, the agitation of the first reactor contributes to the
control of the rubber phase and its average particle size
(2, 28). Tube sizing and spacing are apparently critical in
avoiding excessive local melt temperatures. Computer simulations
have been developed in the study of this type problem (29).
An "eggbeater" type reactor has been shown as a second reactor
and is illustrated in Fig. 18. In this design there are two
hollow agitators with overlapping arms. Heat transfer fluid
circulates through the shaft and arms of these agitators as well
as through the reactor jacket. This double-shafted design was
developed to prevent channeling and rotation of the viscous
polymer melt (2). The reactors are usually divided into a
series of separate coolant circulation zones as illustrated in
Fig. 17 to help provide the desired axial temperature profile.

        Ethylbenzene, generally in 6- 15% concentration
(2, 26, 27) is used as a diluent with the feed, most likely to
reduce viscosity and thereby facilitate melt pumping and heat
transfer. It also serves as a chain transfer agent (30).
With PS (26), the first reactor is maintained between 85 and
130°C. The reacting mass is provided with gentle non-turbulent
stirring to suppress channeling and promote stratification into
layers which increase in conversion as they are slowly forwarded
through the vessel. The syrup exiting the first reactor contains
25 to 50% polymer and enters a second reactor operated on similar
principles, exiting at temperatures up to 150°C and with up to
70% polymer. The discharge from this reactor then enters the
third reactor, also similarly operating, where it is raised to
temperatures preferably between 165 to 185°C where most of the
remaining monomer is polymerized. The exiting melt passes
through a heater raising its temperature to 220 to 240°C, and
then enters the finishing zone which is a vacuum chamber where
most of the ethylbenzene and unreacted styrene are devolatilized.
The devolatilized melt is then withdrawn continuously from the
finishing zone.

        It is likely that some backmixing occurs, especially
in the first reactor where fluid viscosities are relatively low.
As polymerization proceeds and viscosity increases, the
stratified layer condition cited above is gradually approached.

        In process variants for HIPS (27, 28), the feed
solution to the first reactor, besides styrene and ethylbenzene,
will also contain dissolved polybutadiene rubber along with
antioxidants, chain transfer agents, and possibly mineral oil.
The rubber phase particles are formed in the first reactor and
their average size is also largely determined by conditions
existing there. The Ruffing et al patent (27) implies that the
first reactor operates significantly backmixed at temperatures
between 85 and 130°C with sufficient agitation to maintain the
rubber phase uniformly dispersed with a 2-to 25-micron particle

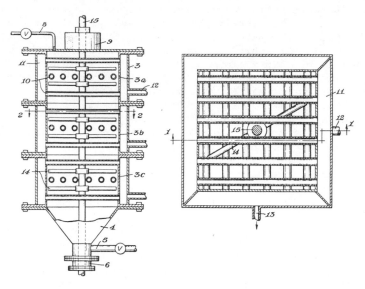

Figure 17.   Dow Chemical Co. patented stratifying polymerization reactor (26)

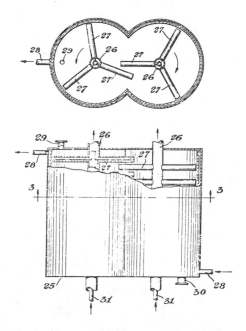

Figure 18.   Dow Chemical Co. patented "eggbeater" agitated tower reactor (27)

size. The exiting polymer content is up to 35%. The claims
indicate that agitation in the downstream reactors need merely
be sufficient to control reaction temperature and macrogel
formation without inhibiting grafting or forming high molecular
weight polymer. In a claim where the syrup exiting the first
reaction contains 7 to 15% polymer, the agitation criteria of
the second reactor are similar to the first.

      An example based on the configuration of Fig. 16 has
2200 lbs./hr. feed with 5.5% rubber entering the first reactor,
described as a Pfaudler reactor of 7500 lbs. operating capacity
with an S-shaped blade stirrer rotating at 67 rpm (27). The
reactor operates at 112°C, and the effluent contains 11% polymer.
This is very close to the phase inversion condition. The syrup
is then fed to the second reactor which is of the "eggbeater"
design, Fig. 18. Syrup flowing through the second reactor
progressively heats up, exiting at 133°C, and containing about
37% polymer. The effluent enters the top of the third reactor.
Although cooling features are not cited, the design is probably
similar to that shown in Fig. 17. The melt temperature progressively increases from 140 to 165° or more, and the effluent
which contains 80% or more of polymer passes through a tubular
heat exchanger where it leaves at about 235°C. The hot melt
flows into a devolatilizer consisting of a vertical vacuum
tower maintained at about 10 torr. The melt enters the top
of the devolatilizer through a die which forms it into falling
strands about 1/4" in diameter to promote the vaporization of
volatile ingredients. The devolatilized melt at 220 to 235°C
falls into a pool at the bottom of this finishing chamber from
which it is pumped for pelletization.

      Finch et al (28), show three "stratifying polymerizers"
rather than the design combinations described earlier by Ruffing
et al (27). The reactors operate at inlet and outlet temperatures respectively of 120 to 135°C, 135 to 145°C, and 145 to
170°C. The first reactor effluent contains 18 to 20% polystyrene and a portion of this stream is recirculated back to the
reactor inlet such that the inlet stream polystyrene concentration is as high as 13.5%. This recirculation is claimed to
improve rubber phase particle size control and end use properties.

      2.3.6 BASF Patents. A BASF process for HIPS as
described by the patent to Bronstert et al (31) is related
closely to the patent of Ruffing et al (40) discussed above.
The principal difference is that a prepolymerizer CSTR has been
placed ahead of the first reactor. An example cited on a
production scale similar to the example discussed in Section
2.3.5, reveals the prepolymerizer to be a CSTR with a 3-blade
agitator operating at 110°C and 7.5% solids. The first reactor
is a roughly similar CSTR operating at 125°C and 35% solids.
The second and third reactors, corresponding to those described
in Section 2.3.5, operate at 125 to 140°C and 140 to 175°C, with

effluents at 60 and 85% solids respectively. The principal point
is that the total prepolymerizer solids is controlled at 1.1 to
2 times the rubber level but not more than 16%. Higher product
gel levels, better rubber particle size control and improved
end use properties are claimed.

2.3.7 Shell Patents. Several patents, (32, 33, 34)
illustrate Shell's technology for HIPS. In one of these (33)
different reactor designs are employed as illustrated in Fig.19.
An example starts with a solution of approximately 9% cis-1,
4-polybutadiene rubber and 2% mineral oil along with antioxidants
being fed to a turbine agitated CSTR operating at 105 to 110°C
and about 8% conversion. This is just slightly ahead of the
phase inversion point. The effluent syrup flows into a second
CSTR fitted with a scroll agitator, operating at about the
same temperature and 23% conversion. It is likely that these two
CSTR's are employed in series to avoid problems with rubber
phase particle size and gel formation arising from excessive
backmixing of syrups below and above the phase inversion point.
The effluent from this second CSTR flows to a fully backmixed
conical reactor equipped with a very slowly turning helical
ribbon agitator. This reactor operates at 173° and 86%
conversion with a 2.4-hr. residence time and is cooled by
refluxing styrene monomer. The exiting syrup then enters a
tower reactor from which it exits after a 5.4-hour average
residence time at about 97% conversion and 207°C. This tower
reactor is unagitated except for wall scraping a few times each
hour which does not disturb the flow pattern of the reacting
mass or produce significant shear. The slow wall scraping is
probably to prevent accumulation of stagnant melt. The tower
effluent passes to a devolatilizer. Inert diluents in
concentrations up 20% can be used to facilitate mixing, pumping,
and heat transfer.

This process uses three CSTR's followed by an LFR for
finishing. The CSTR designs change to accommodate the changing
mixing and heat transfer requirements as conversion rises.

A recent patent (34) describes a process similar in
several ways to the one above. The initial turbine agitated
CSTR is eliminated. Instead, the feed flows to the helically-
agitated CSTR operating at 25-35% conversion and 125-135°C.
This patent then calls for two or more reflux cooled conical
CSTR's with helical ribbon agitators operating in series in an
"intermediate conversion zone" ranging from 65 to 85% conversion.
The conversion in succeeding reactors in this zone should show
a relative difference of 15-25%. As with the earlier patent,
(33) the effluent from the intermediate zone flows to an adiabatic
tower LFR. The stated purpose of the additional intermediate zone
reactors is to reduce the loss of polystyrene occlusions from
the rubber phase particles which is postulated to occur if there
is an excessive difference in conversion between succeeding

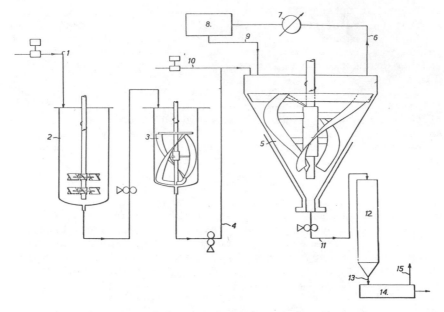

Figure 19. Schematic of Shell patented continuous mass polymerization process
(33)

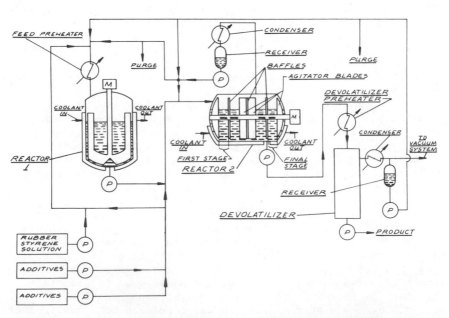

Figure 20. Schematic of Monsanto Co. patented continuous mass HIPS process
(35)

reactors. By avoiding the loss of occlusions, rubber phase volume is retained and product toughness is thereby improved. A similar rationale is the basis of other process reactor designs. (35)

2.3.8 <u>Monsanto Patents</u>. A process for HIPS is described in the patent issued to Carter and Simon (35) and is illustrated in Fig. 20. There are two reactors: an anchor agitated CSTR and a reflux cooled LFR. Both reactors can be operated at variable and controllable fillage so that a given product can be made over a range of rates.

In a typical example (33) a fresh feed of 8% polybutadiene rubber in styrene is added with antioxidant, mineral oil, and recycled monomer to the first reactor at 145 lbs./hr. The reactor is a 100-gallon kettle at approximately 50% fillage with the anchor rotating at 65 rpm. The contents are held at 124°C and about 18% conversion. Cooling is effected via the sensible heat of the feed stream and heat transfer to the reactor jacket. In this reactor the rubber phase particles are formed, their average size determined and much of their morphology established. Particle size is controlled to a large measure by the anchor rpm.

The first reactor effluent flows into the second reactor, an isobaric LFR maintained at about 20 psia. This 50-gallon reactor is about 53" long and operates at about 40% fillage. The agitator is a horizontal shaft on which are set a series of 2" wide paddles alternating at right angles to one another as suggested in Fig. 20. It turns at 15 rpm. Along the shaft and rotating with it are four circular disc baffles with an average axial clearance of about 3/8". These baffles are positioned to divide the reactor into five separate back-mixed stages of approximately equal volume.

The reactor operates with the effluent at about 166°C and 62% conversion. Temperature control is effected primarily by reflux cooling as indicated in Fig. 20 with the condensed vapors being returned to the upstream reactor compartment. Since the syrup solids increase generally stepwise while proceeding from one compartment to the next, and the contents of each compartment are boiling under constant pressure, the temperature in each succeeding compartment increases. It is claimed that the linear flow behavior provided by the reactor staging results in more favorable rubber phase morphology than would be the case if the second reactor were operated as a single CSTR.

The second reactor exit syrup is pumped through a shell and tube heat exchanger which raises its temperature to about 240°C. After this preheating, the melt enters a devolatilizer chamber maintained at 50 torr where volatile materials flash from the melt. The devolatilized melt forms a pool at the base of the vacuum chamber from which it is pumped

through a strand die, cooled and pelletized for further use.
The vapors exiting the devolatilizer are condensed and recycled
to the reactors.

        2.3.9 United Sterling Corporation Patent. The step
of recycling reactor effluent syrups to feed streams as described
earlier in the patent of Finch et al (28) is also found in a
recent patent (36) issued to United Sterling Corporation and
illustrated in Figure 21. Here, a portion of the effluent
at 130° and 35% solids from the second reactor, a horizontal
LFR, is recycled to the first reactor, a CSTR of variable
fillage operating at 98°C and around 14% solids. Solids level
in the first reactor is maintained very close to that of the
phase inversion point. Two LFR's in series downstream of the
second reactor gradually raise the reactant syrup to about 72%
conversion and 170°C. This syrup is then heated to 250°C and
devolatilized in a vacuum chamber maintained at 10 torr.
Improved toughness is claimed for this process.
        By maintaining the first-stage reactor just beyond
the phase inversion point, the dispersed rubber phase is
relatively rich in dissolved styrene. As polymerization
subsequently proceeds in the LFR's, the dissolved styrene will
react to form either a graft copolymer with the rubber or a
homopolymer. The latter will remain within the rubber droplet
as a separate occluded phase. Achieving the first-stage reactor
conversion and temperature by recycling a portion of the hot
second reactor effluent may permit simplification of the first
reactor temperature control system.

        2.3.10 CSTR Designs and Use. A patent granted to
Mitsui Toatsu Chemicals, Inc. (37) describes a styrene
polymerization process involving 3 to 5 CSTR's in series.
It is illustrated in Fig. 22 and shows designs applicable
over a wide range of syrup viscosities.
        The first-stage CSTR is completely filled and
utilizes a pair of opposing axial flow turbines on the agitator
shaft. The maximum recommended viscosity for this reactor is
40 poise and the turbine agitation is set such that total flow
volume generated per unit time by the turbines is 500 to 1000
times the feed rate to the reactor.
        Temperature control is primarily obtained via the
sensible heat of the cooled feed stream with the remaining
heat of reaction being removed by the reactor jacket.
        The remaining reactors are of similar design as shown
in Fig. 22. These are bottom fed, completely filled vessels.
There is a central upward pumping screw surrounded by a draft
tube through which coolant circulates. The reactant syrup
descends in the annular space between the draft tube and the
jacketed reactor wall. In this annular space is a circular
rank of manifolded vertical tubes with circulating coolant

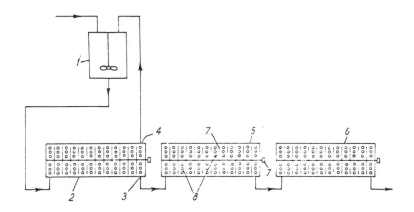

Figure 21. Schematic of United Sterling's patented process for continuous mass HIPS (36)

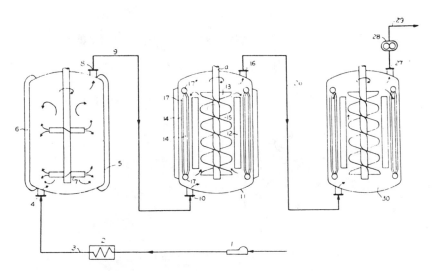

Figure 22. Schematic of Mitsui Toatsu patented continuous mass process for polystyrene (37)

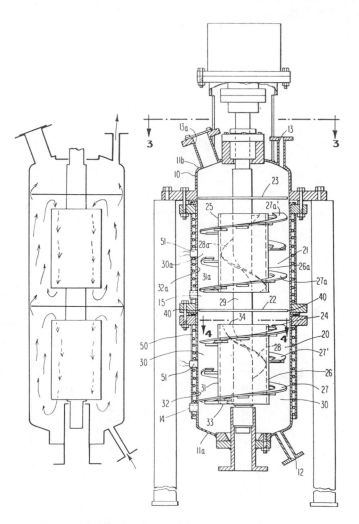

Figure 23.   Patented staged CSTR reactor design (38)

to provide more heat transfer surface. The volumetric pumping rate of the screw is 50 to 100 times the feed rate to the reactor. The high pumping rate and large heat transfer surface are claimed to provide very thorough mixing and top-to-bottom temperature gradients of less than $1^{\circ}C$.

A recent patent (38) describes a polymerization reactor consisting of staged CSTR's and is illustrated in Figure 23. The design is contemplated for viscosities above 100,000 cp. Figure 23 shows a two-stage vertical design and the fluid flow patterns within each stage as well as through the reactor. Each stage has a helical screw mounted within a draft tube set on the central rotating shaft. Rotating in the annular space between the draft tube and the inside vessel wall is an interrupted ribbon agitator having a pitch opposite to that of the helical screw. A rotating disc baffle with a small annular clearance impedes backmixing between stages. The relative pitches and dimensions of the ribbon agitator and helical screw are designed so that, in conjunction with the interstage baffle, draft tube and vessel wall, predetermined portions of the fluid are respectively recirculated within the stage and forwarded to the next stage. The reactor can also be operated horizontally, in which case the ribbon agitator promotes vapor disengagement where evaporative cooling is employed.

Hamielec and coworkers (41, 42, 43) have conducted extensive experimental and theoretical studies with styrene polymerization in CSTR's. Theirs represent probably the first published work in this area at commercially interesting temperatures and conversions relating theory to experiment, and determining the effects of reactor configuration and conditions on conversion, molecular weight and MWD.

2.3.11 Tubular Reactors. Although tubular reactors have received a great deal of theoretical study for over twenty years, they have not generally been favored on an industrial scale where significant heat removal from a viscous polymerizing syrup is required. Under cooling conditions, a very viscous syrup layer forms at the tube wall which can eventually lead to tube blockage (15). A large scale tubular reactor usually consists of many tubes in parallel surrounded by coolant. Flow through parallel cooled tubes is usually unstable and unpredictable. Tubular or unagitated tower reactor designs with PS and HIPS polymerization processes have therefore been used principally as finishing reactors where cooling requirements are small or absent (22, 24, 32, 33).

Experimental work with styrene in tubular reactors has been reported (39) where viscosities were relatively low due to conversions below 32%. However, Lynn (40) has concluded that a laminar flow tubular reactor for styrene polymerization is probably technically infeasible due to the distortion in velocity profile which develops and the resultant large variation in residence time.

Literature Cited

1. Bishop,R.B., "Practical Polymerization for Polystyrene", Cahers, Boston, 1971.

2. Amos, J.L., Polym. Eng. Sci. (1974) 14, (1), 1-11.

3. Biesenberger, J.A. et al.,Appl. Polym. Symp. (1975) 26, 211.

4. Biesenberger, J.A. and Capinpin, R., Polym. Eng. Sci. (1974) 14, 737.

5. Zimmerman, R.L., et al., Adv. Chem. Ser. (1962) 34, 225.

6. Metzner, A.B., A.I.Ch.E.J. (1960) 6, 109.

7. Kern, D.Q., "Process Heat Transfer", Mc Graw-Hill, New York 1950.

8. Ide, Y. and White, J.L., J. Appl. Polym. Sci. (1974) 18, 2997.

9. Uhl, V.W. and Voznick, H.P., Chem. Eng. Prog. (1960) 56 (3),72.

10. Coyle, C.K. et al., Can. J. Ch.E. (1970) 48, 275.

11. Sawinsky, J. et al., Chem. Eng. Sci. (1976) 31, 507.

12. Peters, D.C. and Smith, J.M., Trans.Instn. Chem.Engrs. (1967) 45, 360.

13. Peters, D.C. and Smith, J.M., Can.J.Ch.E. (1969) 47, 268.

14. Smith, J.M., Chem.Eng. (London) (May 1972), 182.

15. Brasie,W.C., "Some Design Considerations for Reactors for Mass and Solution Polymerization". Paper presented at 63rd. National A.I.Ch.E. Meeting, St. Louis, (Feb. 1968).

16. Fong, W.S., Stanford Research Institute Report, 39A (1974), 163.

17. Brit. Pat. 513,256 (to Bakelite Corp.), October 9, 1939.

18. Stein, A. and Walter, R.L., U.S. Pat. 2,886,553 (to Monsanto), May 12, 1959.

19. Simon, R.H.M. and Alford, G.H., Appl. Polym. Symp. (1975) 26, 31-37.

20. Beckman, G., Adv. Chem. Ser. (1973) 128, 37.

21. Albright, L.F. and Bild, C.G., Chem. Eng. (September 15,1975) 82, 121.

22. German Pat. 634,278 (to I.G. Farbenindustrie) 1936.

23. DeBell, J.M. et al., "German Plastics Practice", DeBell and Richardson, Springfield, Ma. (1946) 26-39.

24. Allen, I., et al., U.S. Pat. 2,496,653 (to Union Carbide), Feb. 7, 1950.

25. Albright, L.F., "Processes for Major Addition- Type Plastics and Their Monomers", McGraw-Hill, New York (1971), 328.

26. McDonald, D.L. et al., U.S. Pat. 2,727,884 (to Dow Chemical Co.), Dec. 20, 1955.

27. Ruffing, N.R. et al., U.S. Pat. 3,243,481 (to Dow Chemical Co.), March 29, 1966.

28. Finch, C.R. et al., U.S. Pat. 3,660,535 (to Dow Chemical Co.), May 2, 1972.

29. Brasie, W.C., Adv. Chem. Ser. (1972) 109, 101-105.

30. Jones, C. et al., U.S. Pat. 2,739,142 (to Dow Chemical Co.), March 20, 1956.

31. Bronstert, K. et al., U.S. Pat. 3,658,946 (to BASF), April 25, 1972.

32. Brit. Pat. 1,175,261 (to Shell), Dec. 23, 1969.

33. Brit. Pat. 1,175,262 (to Shell), Dec. 23, 1969.

34. Gawne, G. et al., U.S. Pat. 4,011,284 (to Shell Oil Co.), March 8, 1977.

35. Carter, D.E. & Simon, R.H.M., U.S.Pat. 3,903,202 (to Monsanto Co.), Sept. 2, 1975.

36. Belgian Pat. 75848 (to United Sterling Corp.), Sept. 26,1977.

37. Can.Pat. 864,047 (to Mitsui Toatsu Chemicals,Inc.),Feb.16, 1971.

38. Weber, A.P., U.S. Pat. 4,007,016 (to Bethlehem Corp.), Feb. 8, 1977.

39. Wallis, J.A. et al., A.I.Ch.E.J. (1975) 21, 686-698.

40. Lynn, S., A.I.Ch.E.J. (1977) 23, 387-389.

41. Hui, A.W.T. and Hamielec, A.E., J.Polym.Sci., (1968) C25,167.

42. Duerksen, J.H., A.I.Ch.E.J., (1967) 13, 1807.

43. Duerksen,J.H. and Hamielec,A.E., J.Polym.Sci.(1968) C25,155.

RECEIVED January 12, 1979.

Continuous-Emulsion Polymerization of Styrene in a Tubular Reactor

A. L. ROLLIN, W. IAN PATTERSON, J. ARCHAMBAULT, and P. BATAILLE

Chemical Engineering Department, Ecole Polytechnique de Montreal, C.P. 6079, Succursale "A", Montreal, Quebec, H3C 3A7, Canada

The advantages of continuous tubular reactors are well known. They include the elimination of batch to batch variations, a large heat transfer area and minimal handling of chemical products. Despite these advantages there are no reported commercial instances of emulsion polymerizations done in a tubular reactor; instead the continuous emulsion process has been realized in series-connected stirred tank reactors ($\underline{1}$, $\underline{2}$, $\underline{3}$). A few workers have examined the continuous emulsion polymerization process in a tubular reactor ($\underline{4}$, $\underline{5}$, $\underline{6}$), the initial work being done in the turbulent regime. This flow condition was chosen to maximise heat transfer and mixing, however it was found that a pre-coagulum was formed and resulted in plugging of the reactor ($\underline{4}$). Ghosh and Forsyth (6) examined the emulsion polymerization of styrene in a continuous tubular reactor. They restricted operation to the laminar flow regime and also encountered reactor plugging except when high soap concentrations were employed. They attributed the plugging to emulsion instability and the flow restrictions caused by the thermocouple wells and they also concluded that the kinetics were essentially those of the Smith-Ewart model. Recently, Rollin et al examined the effect of the flow regime on the emulsion polymerization of styrene in a tubular reactor ($\underline{7}$). It was found that the rate of polymerization was a maximum at the laminar-turbulent transition when an "emulsion Reynolds number, $(N_{Re})_e$ is defined as for a circular tube, except that the fluid properties, particularly the viscosity, are those of the emulsion before any appreciable reaction has occurred. (Previous workers had calculated their values of N_{Re} using the viscosity of the latex product). This result was interpreted in the light of work done on the effect of stirring speed (degree of agitation) for batch emulsion polymerizations ($\underline{8}$, $\underline{9}$, $\underline{10}$, $\underline{11}$) where it was found that an optimum stirring speed existed. This optimum was first explained by Evans et al ($\underline{8}$). It may be interpreted for tubular reactors at low

'Current address: Hydro-Quebec, Montreal
0-8412-0506-x/79/47-104-113$06.00/0

Reynolds numbers as follows: in the laminar flow regime the reaction is diffusion controlled although the velocity gradient aids the diffusion. Very small values of N_{Re} permit phase separation to occur thus greatly diminishing the area available for monomer transfer. Highly turbulent flow (large values of $(N_{Re})_e$) promotes the break-up of monomer drops thus reducing the soap available to form micelles and hence the number of polymer particles is reduced. Since the rate of polymerization (Smith-Ewart) is given by:

$$r_p = (k_p\ M_p\ N_t)/2 \qquad\qquad (1)$$

where N_t = the number of polymer particles per unit volume it is expected that the rate of polymerization will diminish due to the decreased value of N_t.

Two additional results were noted during this series of experiments. It was found that plugging of the reactor occurred when the conversion reached about 60%. No satisfactory explanation or cause for the plugging was determined. It was also noted that, regardless of the rate of polymerization, no further reaction occurred after a period of about 60 to 75 minutes. This is in contrast to reaction times of up to three hours for the same recipe used in a batch reactor.

Effect of Reactor Geometry

The studies described above have all been done in essentially straight, tubular reactors. The curved part (the elbows) constituted a very small proportion of the total length. A reactor of commercial interest operated at an $(N_{Re})_e$ in the vicinity of 2100 would be on the order of 60 metres to 600 metres in length depending on tube diameter. The advantages of good heat transfer and temperature control of emulsion polymerization are generally realized by the immersion of the reactor in a heat transfer medium of substantial thermal capacity. This is difficult and costly to achieve when a linear tubular reactor is used and a helical configuration suggests itself as a viable alternative. The emulsion polymerization of styrene in a straight tubular reactor has been shown to be sensitive to the hydrodynamics of the flow. Thus it is reasonable to ask what the effect of a helical reactor geometry would be.

The flow through helically configured, round tubes was first examined by Eustice (12) in 1910 and the first theoretical analysis was published by Dean in 1927 and 1928 (13). He showed that the fluid flow could be characterized by the dimensionless group

$$(N_{Re})^2\ \frac{d_i}{D} = (N_{Dn})^2$$

where: d_i = the tube inside diameter.
        D = the diameter of the helical bend.
        N_{Dn} = the Dean number

N_{Re} = the conventional Reynolds number,
but was unable to account for the secondary flow observed by
Eustice. White (14) derived an expression for the laminar flow
pressure drop in a helix and Taylor (15) later showed that the
effect of tube curvature was to damp flow perturbations and
increase the value of the Reynolds number at which the laminar-
turbulent transition occurs. Since then a number of workers
have derived various expressions to predict the transition and
the flow characteristics in the different regimes. This work is
summarized by Srinivasan et al (16).

The conflicting predictions of the various equations for the
transition point have led us to experimentally determine the
laminar-turbulent transition for the particular configuration
employed in this work. This is reported in the section on
results.

Experimental

The emulsion polymerization of styrene was carried out in an
open loop reactor shown schematically in Figure 1. The apparatus
consisted of an emulsification circuit connected to a tubular
reactor of a 2.23 cm i.d. fluorinated polymer tube 155.7 m. long
in a helically coiled configuration. The diameter, the number of
turns and the tube length of each coil are presented in Table I.
It is observed that the tube length, or number of loops, of each
of the four coils is progressively longer. The connections
between the coils are straight lengths of stainless pipe with
taps to permit sampling of the solution at the points S_2 to S_6.

An agitated vessel containing the monomer, emulsifier and
water was used to supply the mixture to a sonic emulsifier
(Sonolator A from Sonic Eng. Ltd) fed by a gear pump. A return
line was installed to regulate the flow of the stabilized emulsion through the reactor. The aqueous solution of the initiator
was continously injected by a peristaltic pump (Masterflex model
7016) upstream of a restriction at the entrance to the heating
section. Instantaneous mixing of the initiator in the emulsion
was achieved by the energetic eddies created by the restriction
(0.63 cm diameter). The straight stainless steel heating section
(2.54 cm i.d.) was jacketed by a 6.3 cm i.d. galvanized pipe
and counter-currently flowing water at 62°C circulated in the
jacket to maintain the desired emulsion temperature of 60°C at
the exit of this section.

The reactor was operated in the following manner. First,
the required volumes of the emulsion ingredients were placed in
the agitated reservoir and the operating temperatures were
established in the heat exchanger, the reactor coil tanks and
the heating jacket of the agitated vessel. A nitrogen blanket
was injected at the top of the reservoir to avoid oxygen absorption and the agitator was started (constant speed of 110 rpm).
Emulsification was achieved by circulating the reservoir mixture

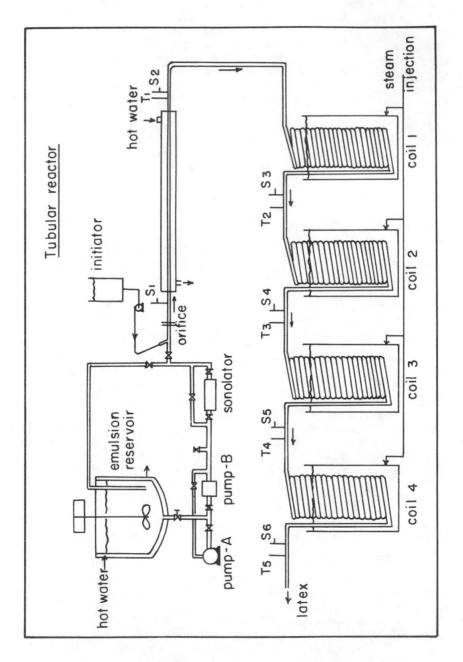

Figure 1.   Schematic of the helically coiled tubular reactor

TABLE I: REACTOR DIMENSIONS

Section	Length (m)	Tube diameter (cm)	Cumulative length (m)	Cumulative volume (cm^3)	Coil Diameter (cm)	Coil Number of loops
Mixing of initiator	0.59 0.12	0.63 2.54	0.59 0.71	107	–	–
Heat Exchanger	3.45	2.54	4.17	2,027	–	–
Coil # 1	21.70	2.23	27.57	11,087	43.2	16
Coil # II	32.55	2.23	61.91	24,387	43.2	24
Coil # III	37.98	2.23	102.03	40,187	43.2	28
Coil # IV	54.26	2.23	159.98	62,487	43.2	40

through the gear pump which was monitored for safe operating conditions. The closed loop recirculation via the pump and Sonolator for a period of approximately 75 minutes provided a very stable emulsion.

When the emulsification was complete, the gear pump speed was reduced to the desired level and the flow rates of the emulsion and of the initiator were established. Temperature profiles and pressure drops along the reactor were recorded and the flow rate was measured gravimetrically.

Fifteen experimental runs were carried out in a range of emulsion Reynolds numbers from 1350 to 10600 at a constant temperature of 60°C. The duration of each run was approximately three (3) residence times. All runs used the emulsification formulation given in Table II. The five thermistors placed along the reactor verified that no temperature gradient existed during a run and the range of flow rates varied from 0.64 to 5.17 Kg/min. Samples were collected at fixed times from the sample points and placed in closed test tubes containing a small amount of hydroquinone inhibitor. The styrene conversion was gravimetrically determined.

TABLE II: EMULSION FORMULATION

Workers	$[S] \times 10^{-3}$ (mol/1)	Potassium Persulfate $[I] \times 10^{-3}$ (mol/1)	Styrene (mol/1)
Omi (9)	5.88*	2.42	1.69
Rollin (7)	6.52**	2.55	1.74
B-3 & B-4 (this work)	12.60**	3.07	1.74

Note: concentrations referred to unit volume of emulsion

*   Oleate Sodium

**   Dodecyl Sodium Sulfate

Molecular weights were measured using a Waters gel permeation chromatograph (model 200). Complete details of the equipment and procedures can be found in references (17, 18, 19).

Results and Discussion

A summary of the nine batch reactor emulsion polymerizations and fifteen tubular reactor emulsion polymerizations are presented in Tables III & IV. Also, many tubular reactor pressure drop measurements were performed at different Reynolds numbers using distilled water to determined the laminar-turbulent transitional flow regime.

Pressure drop measurements: The pressure drop in a coiled tube configuration is predicted to be higher than in a straight tubular reactor and the laminar-turbulent transition is also predicted to be shifted to higher Reynolds numbers because of the presence of the centrifugal force acting on the fluid elements. Shown in Figure 2 are curves representing the hydraulic behavior in straight tube and in a coil of dimensions identical to that used in the experimental reactor. Curves 1 and 2 were calculated using Ito's equation (20) and the curves D are the well known Moody curves for straight smooth circular tubes. The experimental data, represented by the points in Figure 2, fall on the upper curves (1 & 2) confirming that the measured pressure drops correspond to a coiled configuration rather than a straight tube.

White (14) proposed a graphical method of determining the critical Reynolds number N_{Rec} at which the fully turbulent flow exists in a coiled tube. As shown in Figure 3, the curve representing White's equation for the ratio of the logarithm of the friction factor in a coil (fc) to the friction factor in a straight tube (f) versus the logarithm of the Dean number, N_{Dn}, fits the experimental data well for Reynolds numbers smaller than approximately 5500. The intersection of the straight line passing through the experimental points at high Reynolds numbers with White's curve represents the flow condition at which fully turbulent flow occurs. It can be observed from Figure 3 that the intersection occurs in the region $5500 \leq Nre \leq 7100$ compared to a value of $N_{Rec} = 7830$ calculated using Holland's equation (16). The difference between the calculated and experimentally determined critical Reynolds number can be explained from the reactor configuration, which consisted of four coils connected by straight tubing section. The straight sections would lower the N_{Rec} relative to the value for a single coil. It is justified to note that higher polymerization rates are forecasted for Reynolds numbers in the vicinity of 5000 to 7000.

Batch Polymerizations: Nine batch polymerizations were performed to verify that our formulation behaviour was unchanged from

TABLE III: SUMMARY OF EXPERIMENTAL BATCH POLYMERIZATIONS

Run	Emulsion Formulation (mol/l emulsion)				Emulsification			Reaction	
	$S \times 10^{-3}$ (mol/l)	$I \times 10^{-3}$ (mol/l)	Styrene M	D*/I	Volume (l)	Time (min)	Agitation speed	Volume (l)	Agitation speed (rpm)
B-1	5.44	1.77	1.33	D	1.0	30	650 20 rpm	Apparatus C-1 1.0	650
B-2	12.60	3.07	1.74	D	3.0	20	Apparatus E-1	"	650
B-3	12.60	3.07	1.74	D	1.0	45	670 ± 20 rpm	"	670
B-4	12.60	3.07	1.74	D	1.0	45	670 ± 20 rpm	"	670
B-5	12.60	3.07	1.74	D	1.0	45	670 ± 20 rpm	"	550
B-6	12.60	3.07	1.74	D	1.0	45	670 ± 20 rpm	"	930
B-7	12.50	3.00	1.74	D	200.0	30	Apparatus E-2	"	670
B-8	11.96	2.96	1.74	D	200.0	60		Apparatus C-2 0.73	670
B-9	12.20	3.06	1.74	I	195.0	75			670

* D = distilled Styrene and I = inhibited styrene

TABLE IV: SUMMARY OF EXPERIMENTAL TUBULAR POLYMERIZATIONS

Run	Flow rate (Kg/min)	%V (1)	$(N_{Re})_c$ (2)	Emulsion formulation (mol/l emulsion)				Volume emulsion (l)	Temperature (°C)
				$s \times 10^3$	$I \times 10^3$	Styrene M	Styrene D/I		
T-1	2.52	5.2	5170	12.60	2.92	1.76	I	190	60 ± 2
T-2	0.80	6.2	1650	12.44	3.68	1.72	I	205	60 ± 2
T-3	0.91	5.4	1870	6.50	3.24	1.74	I	205	60 ± 1
T-4	3.10	3.0	6370	12.01	2.81	1.66	D	230	62 ± 2
T-5	0.64	8.6	1330	12.54	3.20	1.74	I	185	60 ± 3
T-6	1.44	6.6	2900	12.57	3.09	1.74	D	212	62 ± 2
T-7	1.59	5.9	3260	12.02	2.74	1.66	I	190	61 ± 2 (73*)
T-8	0.86	5.7	1770	13.32	6.81	1.84	I	205	61 ± 2
T-9	1.80	5.2	3700	11.93	3.23	1.65	I	225	62 ± 2 (71*)
T-10	0.90	5.5	1860	12.50	3.27	1.74	I	205	61 ± 1
T-11	0.65	7.5	1350	12.60	2.91	1.75	I	195	61 ± 2
T-12	5.17	2.5	10600	12.60	2.79	1.75	(3)	221	63 ± 3
T-13	4.39	2.9	9010	12.57	2.97	1.75	I	205	61 ± 2
T-14	1.38	6.7	2850	11.91	3.23	1.65	I	221	60 ± 2 (69*)
T-15	4.00	3.2	8210	12.63	2.80	1.74	I	221	63 ± 2

(1) Initiator volumetric flow rate as percent of total flow rate.
(2) Emulsion Reynolds number
(3) 60% distilled styrene + 40% inhibited styrene
* At sampling point S_5

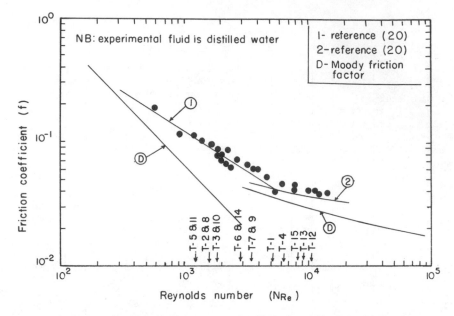

Figure 2.   *Friction factors measured (with water) in the helical reactor*

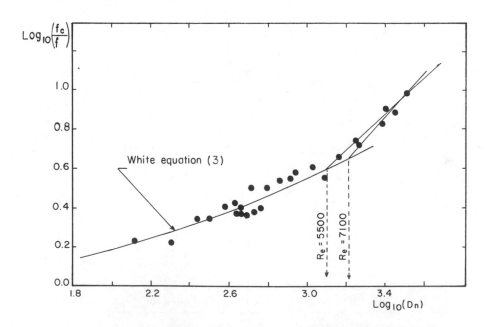

Figure 3.   *Log of friction factor ratio vs. log D_n used to determine the laminar-turbulent transition in the reactor*

previous experiments used to determine the influence of the agitation speed upon the reaction. These polymerizations also served to establish the emulsification time of 75 minutes. The conversion versus the reaction time results for runs B-3 to B-6 (Figure 4) show that both the final conversions and the rates of polymerization for runs at which the agitation speed was 670 rpm were higher than for runs B-5 and B-6. The comparison of our formulation to that of previous workers can be observed on Figure 5 where the conversion versus reaction time results for runs B-3 and B-4 are compared to the results of Omi (21) and Rollin (7) at similar agitator speeds. The conversion rates obtained during this experiment were higher than those obtained by Omi (21) and Rollin (7). This behavior is explained by the higher emulsifier and initiator concentrations in this formulation. The data of runs B-3 and B-4 showed very good reproducibility of the experimental and analytical methods.

The emulsification time required to produce a stable emulsion using the sonic emulsifier was determined experimentally. Figure 6 gives the conversion of styrene versus the reaction time of runs B-2, B-7, B-8 and B-9 and a slight difference is noticed for the longer emulsification time of 75 minutes (which used an operating pressure of 6.88×10^5 N/m^2). On the other hand, the emulsifier operating pressure for run B-2 was slightly higher, being 2.06×10^6 N/m^2, increasing the emulsification yield. An operating pressure of 6.88×10^5 N/m^2 was chosen as the maximum safe pressure and a 75 minute emulsification time was adopted for all runs in the tubular reactor.

Continuous Polymerizations: As previously mentioned, fifteen continuous polymerizations in the tubular reactor were performed at different flow rates (i.e. $(N_{Re})_e$) with twelve runs using identical formulations and three runs having different emulsifier and initiator concentrations. A summary of the experimental runs is presented in Table IV and the styrene conversion vs reaction time data are presented graphically in Figures 7 to 9. It is important to note that the measurements of pressure and temperature profiles, flow rate and the latex properties indicated that steady state operation was reached after a period corresponding to twice the residence time in the tubular reactor. This agrees with Ghosh's results (6).

The styrene conversion versus reaction time results for runs in the laminar flow regime are plotted in Figure 8. Both the rate of polymerization and the styrene conversion increase with increasing flow rate as noted previously (7). The conversion profile for the batch experimental run (B-3) is presented as a dashed line for comparison. It can be seen that the polymerization rates for runs with $(N_{Re})_e > 2850$ are greater than the corresponding batch polymerization with a conversion plateau being reached after about thirty minutes of reaction. This behavior is similar to the results obtained in a closed loop tubular reactor (7) and is probably due to an excessively rapid consumption of initiator in a

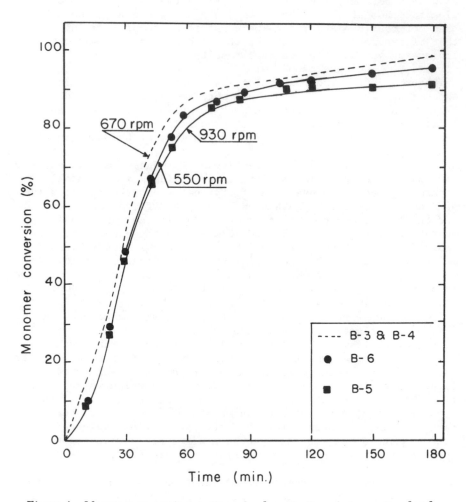

Figure 4.   *Monomer conversion vs. time of polymerization of styrene in a batch reactor: agitation speed as parameter*

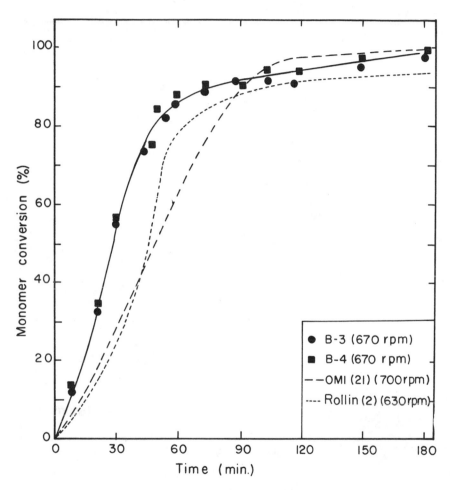

Figure 5. Monomer conversion vs. time of polymerization of styrene in a batch reactor: reproducibility check

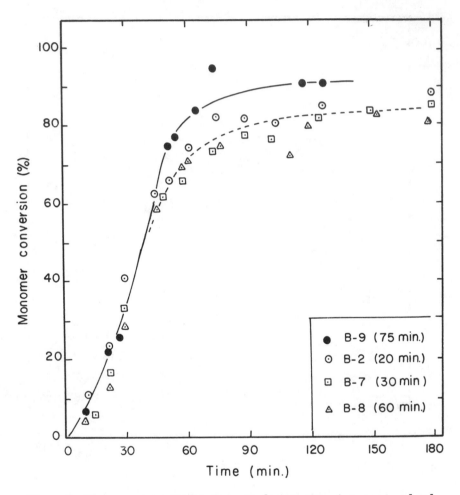

Figure 6.   Monomer conversion vs. time of polymerization of styrene in a batch
reactor: emulsification time as parameter

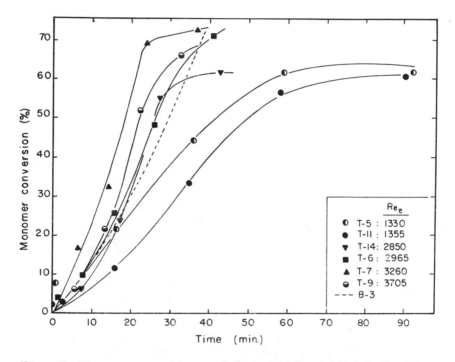

Figure 7. Monomer conversion vs. polymerization time in the helical tubular reactor: laminar flow regime

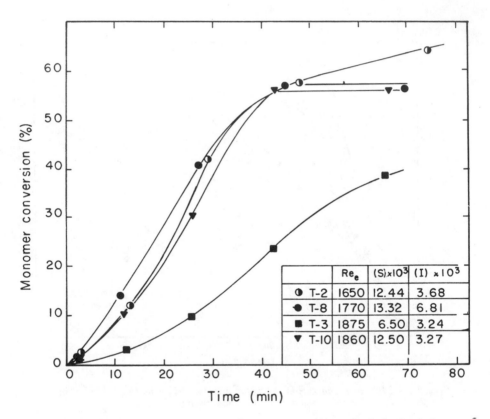

Figure 8.   Monomer conversion vs. polymerization time in the helical tubular reactor: effect of varying initiator and emulsifier concentrations

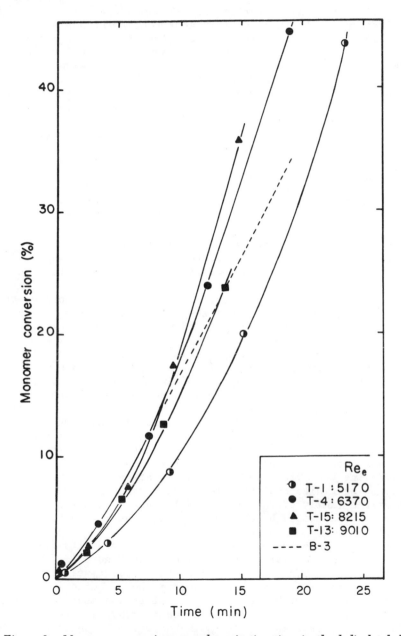

Figure 9.   Monomer conversion vs. polymerization time in the helical tubular reactor: transition and turbulent flow regimes

tubular reactor. The conversion profiles for the runs performed
with different emulsifier and initiator concentrations are presented on Figure 8 together with the profiles of runs T-2 and T-10
for comparison. It is seen that the conversion rate for run T-3
is lower than all the other runs performed at equivalent Reynolds
number. This result agrees with previous work (22) indicating less
conversion with a decrease in the emulsifier concentration. The
results for run T-8 are in agreement with the increased initiator
concentration used for this run.

Finally, monomer conversion profiles for runs performed at
higher Reynolds numbers, corresponding to the turbulent flow
regime, are plotted in Figure 9. Even though the residence times
in the reactor of these runs preclude reaction times greater than
twenty-five minutes, the monomer conversion rates during the propagation period can be estimated. The flow agitation effect on the
polymerization rate versus the emulsion Reynolds number curve is
shown in Figure 10.

It is evident that the shift of the laminar-turbulent transition for the helically coiled reactor corresponds approximately to
the observed shift in the maximum of the rate of conversion curve.
Values of $(N_{Re})_e$ less than 200 give rates of conversion (r_p) that
are roughly comparable to, but slightly larger than those of the
straight tubular reactor. Since the rate is interpreted as being
flow-aided diffusion controlled in this region, any flow distur-
bance will tend to increase r_p. The straight tube flow has less
damping and is thus more susceptible to pertubations and it would
be expected to have a slightly higher rate of conversion. The
values of r_p for $2000 \leqslant (N_{Re})_e \leqslant 6000$ in the helical reactor continue to increase with increasing r_p, and are larger than the
rates observed in the linear reactor. We propose that the rate
is still determined by flow-aided diffusion in this region. The
higher rates are explained by the increased velocity gradient and
the appearance of the secondary flows. In the fully turbulent
flow regime monomer drops are broken (or prevented from coalescing)
this reducing the effective soap concentration, hence the number
of polymer particles and the rate of conversion is lower. Figure
11 shows the family of r_p versus $(N_{Re})_e$ curves as a function of
$(\frac{di}{D})$ or N_{Dn} that would be obtained if this interpretation is correct.

Molecular weights were determined and the results for the
weight average molecular weight are shown in Figure 12 for conversions of 30% and 50%. It is noted that molecular weight decreases
with increasing $(N_{Re})_e$ until just before the turbulence occurs.
This is consistant with the flow-aided diffusion condition postulated for this region interpreted as an increase in the effective
initiator concentration as $(N_{Re})_e$ increases. The increased initiator concentration causes the reduction in molecular weight (22).
Values of $(N_{Re})_e$ greater than 4000 show an increasing weight as
$(N_{Re})_e$ increases. This cannot be explained by the decrease in
effective soap concentration because this would cause a further

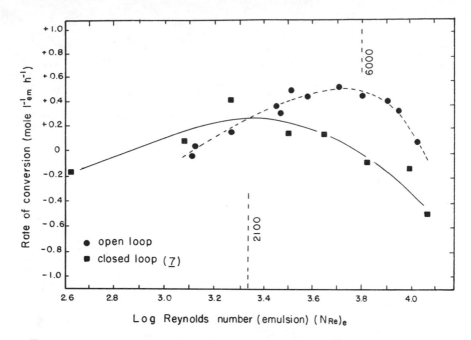

Figure 10.   *Monomer conversion rates as a function of emulsion Reynolds number for straight and helical tubular reactors*

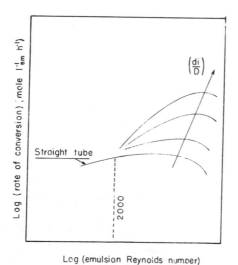

Figure 11.   *Hypothetical conversion as a function of Reynolds number in helically coiled tubular reactors*

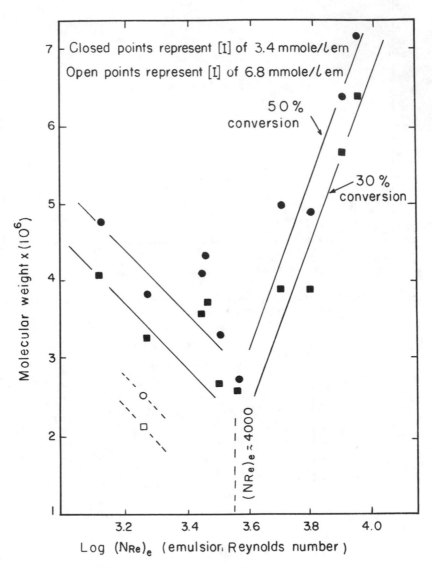

Figure 12.   Molecular weight vs. emulsion Reynolds number at conversion rates
of 30% and 50%

decrease in molecular weight contrary to the observed trend. It may be that turbulent flow in a tubular reactor causes a decrease in the effective initiator concentration as the turbulence becomes more pronounced. This would also be consistant with a decrease in r_p in the turbulent flow region. Unfortunately we have no evidence to corroborate this hypothesis. Number average molecular weights (Mn) were calculated and found to be approximately independent of flow regime but the data had considerable scatter. Particle size distributions were measured but were found to be invalid due to an error in experimental technique. The data reported for the straight tubular reactor (7) showed a broadening size distribution as $(N_{Re})_e$ increased which lends some credence to the idea of a decreasing effective initiator (or radical) concentration with increasing turbulence. The exact role of the persulfate initiator is still in dispute (23) and thus it must be stated that the proposed hypothesis is quite conjectural, and we are not prepared to advance a mechanism to account for a possible decrease in effective initiator concentration.

Conclusions

The work reported here is part of a continuing program on the emulsion polymerization of styrene in a tubular reactor. It is now evident that the reactor construction is of primary importance in avoiding the problem of reactor plugging. The plugging is associated with a wall effect so that both the reactor dimensions and the nature of the wall surface are important.

The maximum rate of polymerization has been confirmed to occur at the laminar-turbulent flow transition. The rate of polymerization was observed to be maximum at the transition for both straight reactors as well as for the helically-coiled reactor for which the transition is at a Reynolds number higher than that of the straight tube. The helically coiled tubular reactor is of industrial interest since it is much more compact and, consequently, the cost and the temperature control problems are more tractable.

It is noted that the maximum value of r_p in the helically coiled reactor is larger than the maximum observed in the straight tube reactor. The r_p increases with increasing Reynolds number while the molecular weight (at a given conversion) decreases. These observations are consistant with the proposed mechanism of the reaction being diffusion controlled in the laminar flow regime. The mass transport is aided by the velocity gradient and thus the reaction rate increases as the Reynolds number is increased.

The mechanism in the fully turbulent regime is less clear. It was originally proposed (7) that, for polymerizations using soap concentrations near the CMC, the highly turbulent flow either prevented the coalescense of monomer drops or increased their number thus reducing the effective soap concentration.

This, in turn, reduced the number of polymer particles (the loci
of reaction) and hence the reaction rate fell. However, this
explanation is at variance with the results reported in Figure
12 where the molecular weight (weight-average) clearly increases
with increasing Reynolds number. It seems more likely that the
turbulent flow results could be explained by a decrease in the
effective initiator concentration. This low concentration would
also explain why there is no further reaction after a period of
about one hour as contrasted with the batch reactions where the
reaction is still proceeding after two to three hours. The
current absence of corroborating evidence makes this explanation
very tentative.

Acknowledgements

The authors express their appreciation to J. Masonnave for
his support and comments. We are grateful to Monsanto Canada
Limited for the donation of the styrene. This work was funded
in part by The National Research Council of Canada (grant
A-6695) and The Quebec Education Ministry (grant CRP 294-72).

Abstract

A continuous reactor has several advantages compared with
the more conventional batch reactor. The more obvious of these
are the absence of handling and contact with potentially toxic
materials and the uniformity of product made. The emulsion
polymerization process is attractive because of the benefits
accruing from improved temperature control due to good heat
transfer. However, no successful process has been reported to
date, with reactor plugging being a primary problem. Ghosh and
Forsyth (1) operated a tubular reactor for the emulsion polymerization of styrene in a laminar flow regime (N_{Re} = 120). They
were able to avoid reactor blockage only at extremely high soap
concentrations. Rollin et al (2) found that the rate of polymerization of styrene in a tubular reactor was a maximum at the
transition from laminar to turbulent flow ((N_{Re})$_e$ = 2100),((N_{Re})$_e$
being the Reynolds number based on the emulsion rather than latex
properties).
    The work reported here used a tubular reactor of approx. 2.5
cm id and   150 meters in length. The reactor, lined with a
fluorinated polymer, was coiled in a helical shape. The recipe
employed standard concentrations of initiator and emulsifier.
It was found that the maximum rate of polymerization occurred at
(N_{Re})$_e$ 5000. This shift in (N_{Re})$_e$ corresponds to the shift of
the laminar turbulent transition in a helically coiled tube as
reported by White (3). Further, no plugging of this reactor,
under any conditions of operation, was noticed. The reaction
mechanism appears to be very close to the Smith-Ewart model,
although conversions were not always as complete as expected.

Nomenclature

d_i inside diameter of reactor, cm

D diameter of helical bend, cm

k_p polymerization propagation rate constant, h^{-1}

M_p monomer concentration, mol l_{em}^{-1}

N_{Re} Reynolds number, dimensionless

$(N_{Re})_c$ value of the Reynolds number at the laminar - turbulent transition, dimensionless

$(N_{Re})_e$ Reynolds number based on emulsion properties prior to polymerization, dimensionless

N_{Dn} Dean number, dimensionless

r_p rate of polymerization, mol $l_{em}^{-1} h^{-1}$

Note: Apparatus C-1 is a laboratory round-bottom flask and laboratory stirrer

                 C-2 is a laboratory round-bottom flask and laboratory stirrer

                 E-1 is a stainless, cylindrical reservoir, piston pump and sonolator

                 E-2 is the reservoir of the helical reactor, gear pump and sonolator

References

(1)   Ghosh, M. and Forsyth, T.H., ACS Symp. Series, No. 24, paper 24 (1976)

(2)   Rollin, A.L., Patterson, I., Huneault, R. & Bataille, P., Can. J. Chem. Eng., 55, 565 (1977)

(3)   White, C.M., Proc. Royal Soc., A-123, 645 (1929)

Literature cited

(1)   Harada, M., Nomura, M., Kojima, H., Eguchi, W. and Nagata, S., J. Appl. Poly. Sci., 16, 811 (1972)

(2)   U.S. Patent No. 2, 831, 842, Dupont de Nemours & Co.

(3)   De Graff, A.W. and Poehlein, G.W., J. Poly. Sci., A-2, 9, 1955 (1971)

(4)   Feldon, M., McCann, R.F. and Laundrie , R.W., India Rubber World 128, 1 (1953)

(5)   Canadian Patent No. 907795, Gulf Oil Canada Ltd (1972)

(6)   Ghosh, M. and Forsyth, T.H., ACS Symposium Series, No. 24, paper No. 24 (1976)

(7) Rollin, A.L., Patterson, I., Huneault, R., Bataille, P.,
     "The Effect of Flow Regime on the Continuous Emulsion Poly-
     merization of Styrene in a Tubular Reactor", Can. J. of Chem.
     55, 565 (1977)
(8) Evans, C.P., Light, J.D., Marker, L., Santoaiala, A.T. and
     Swetting, O.J., J. Appl. Poly. Sci., 5, 31 (1961)
(9) Omi, S., Shiraishi, Y., Sato, H. and Kubota, H., J. Chem.
     Eng. Japan, 2, 1. 64 (1969)
(10) Nomura, M., Harada, M., Eguchi, W. and Nagata, S., J.
     Applied Poly. Sci., 16, 835 (1972)
(11) Omi, S., Kuwabara, L. and Kubota, H., J. Chem. Eng. Japan,
     6, 343 (1973)
(12) Eustice, J., Proc. Royal Soc. A 84, 107 (1910)
(13) Dean, W.R., Phil. Mag., 4, 208 (1927) and Phil. Mag., 5,
     673 (1928)
(14) White, C.M., "Streamline Flow through Curved Pipes", Proc.
     Royal Soc., A 123, 645 (1929)
(15) Taylor, G.I., "The Criterion for Turbulence in Curved Pipes",
     Proc. Royal Soc., A 124, 243 (1930)
(16) Srinivasan, P.S., Nandapurkar, S.S. & Holland, F.A.,
     "Pressure Drop and Heat Transfer in Coils", The Chem. Engr.
     218, CE 113 (1968)
(17) Blanc, M., M.Sc.A. Thesis Chem. Eng. Dept., Ecole Poly-
     technique, Montreal, (1977)
(18) Huneault, R., M.Sc.A. Thesis Chem. Eng. Dept., Ecole
     Polytechnique, Montreal (1976)
(19) Archambault, J., M.Sc.A. Thesis Chem. Eng. Dept., Ecole
     Polytechnique, Montreal (1977)
(20) Ito, H., "Friction Factors for Turbulent Flow in curved
     Pipes", J. Basic, Eng. Trans. A.S.M.E., D, 81, 123 (1959)
(21) Omi, S., Sato, H. and Kubota, H., J. Chem. Eng. Japan, 2,
     1, 55 (1969)
(22) Gardon, J.L., "Mechanism of Emulsion Polymerisation", AIChE
     Symp., May 1969
(23) Blackley, D.C., "Emulsion Polymerization", Wiley, New York
     (1975)

RECEIVED February 6, 1979.

Polyamidation in the Solid Phase

R. J. GAYMANS and J. SCHUIJER

Twente University of Technology, Department of Chemical Technology,
P.O. Box 217, Enschede, The Netherlands

Polyamides can be polymerized in the solid-phase in an oxygen-free atmosphere at a temperature range of 20° - 160°C below their final melting point (1-9).
The results from the literature are not easy to interpret due to the limited temperature ranges, small variations in particle sizes and the occurrence of side reactions with chain branching.

We studied the polyamidation in the solid phase process of nylon 4,6, which has a high melting transition (264° - 320°C) and does not show any tendency to gel (10).

The rate of the solid phase polymerization (SPP) depends on

- the kinetics of the chemical reaction
- the diffusion of the reactive groups
- the diffusion of the condensate out of the particle
- the diffusion at the particle - gas interface
- the heat transfer

The polymerization rate is controlled by the slowest process. Thus it is important to establish the rate controlling steps. The starting material for the (SPP) can be the dry nylonsalt (3,4) but mostly a low or middle molecular weight polymer is used. The polyamide-salts have the disadvantage of high amine losses (3,4).

Griskey (5) and Chen (6) studied the reaction of nylon 6,6 and 6,10 in a SPP in a stream of dry nitrogen in the temperature range of 90° - 180°C. They found that the reaction limiting step was not the diffusion of water but the chemical reaction. The kinetic relationship they observed was

$$\overline{M}n = k \ t^{n} \qquad (1)$$

n = 0,5 and 1.0 for nylon 6,6 and nylon 6,10, respectively. The activation energies of the rateconstant k are respectively 10.5 - 12.96 and

13.2 kcal.mol.$^{-1}$. Monroe (7), reporting on nylon 66, found a dramatic effect of the starting molecular weight on the reaction rate. Increasing the starting molecular weight by a factor of two decreased the reaction time to reach $\overline{M}_n$ = 15,000 by a factor two. Zimmerman (8), comparing SPP and melt-polymerization, showed that in the presence of water the SPP leads to much higher molecular weights at a given pressure than the melt-polymerization. At the same time he noticed a broadening in the molecular weight distribution (m.w.d.) of nylon 6,6.

Ramsey and Dunnill (9) reported the formation of highly branched structures in nylon 6,6 by reacting under anhydrous conditions. According to them this could be prevented by reacting under a blanket of super-heated steam.

A theoretical study of the m.w.d. broadening during the SPP of a semi-crystaline polymer showed that for linear structures, according to the Schulz-Flory relationship, no narrowing or broadening of the m.w.d. is to be expected (11).

The kinetics of the melt-polymerization of nylon 6,6 is third order (1)

$$- \frac{d[-COOH]}{dt} = k. \ [-COOH] \ \Big([-COOH] \ [-NH_2] - [-COOH]_{eq} \ [-NH_2]_{eq}\Big) \quad (2)$$

The rate constante k has an activation energy of 16.8 kcal.mol^{-1} (12). In water free conditions equation 2 can be simplified to

$$- \frac{d[-COOH]}{dt} = k. \ [-COOH]^2 \ [-HN_2] \quad (3)$$

If the polymer is balanced with [-COOH] = [-NH$_2$] the integrated equation is

$$\Big(\frac{1}{[-COOH]_t}\Big)^2 - \Big(\frac{1}{[-COOH]_o}\Big)^2 = 2 \ kt \quad (4)$$

For an unbalanced polymer with [-COOH] - [-NH$_2$] = D and the assumption that D has a constant value results in the integrated equation

$$\frac{1}{D} \ ln \ \frac{[-COOH]}{[-NH_2]} - \frac{1}{[-COOH]} = Dkt + const. \quad (5)$$

The SPP reaction is not necessarily third order. If the endgroups are unbalanced a good approximation of the SPP reaction order can be obtained by expressing it as function of

$$\sqrt{[-COOH] \ [-NH_2]} = \sqrt{P}$$

The rate of reaction is then

$$- \frac{d[-COOH]}{dt} = k. \ \Big(\sqrt{[-COOH] \ [-NH_2]}\Big)^n = k. \Big(\sqrt{P}\Big)^n \quad (6)$$

The integrated form with the unknown order n is then

$$\left(\frac{1}{\sqrt{P_t}}\right)^{n-1} - \left(\frac{1}{\sqrt{P_o}}\right)^{n-1} = (n-1)kt. \tag{7}$$

If the endgroups are balanced, $\frac{1}{\sqrt{P}} = \overline{M}_n$ the equation is

$$\left(\overline{M}_{n_t}\right)^{n-1} - \left(\overline{M}_{n_o}\right)^{n-1} = (n-1)kt. \tag{8}$$

For evaluation purposes it is changed to

$$\left(\sqrt[a]{\overline{M}_{n_t}^a - \overline{M}_{n_o}^a}\right)^{n-1} = (n-1) kt. \tag{9}$$

where a is chosen as near as possible to (n-1).

We studied the polyamidation of nylon 4,6, and varied the reaction time, reaction temperature, partical size, starting molecular weight, and type of reactor gas. At the same time we looked at the molecular weight broadening and the degradation with colour formation. In order to have good heat and mass transfer the reactions were mainly conducted on fine powder in a fluidized bed reactor and with dry nitrogen as carrier gas.

EXPERIMENTAL

The starting materials were low molecular weight polymers prepared by reacting 1,4 diaminobutane and adipic acid for two hours at 220^oC in a capsule in an autoclave (*10*). The low molecular weight material was powdered by crushing and ballmilling.

The fluidized bed reactions were carried out in a glass reactor (fig. 1) 2.5 cm diameter and 50 cm long. The reactor was heated in an oven in which the temperature could be controlled within 0.2^oC. As carrier gases dry nitrogen and super-heated steam, both at a pressure of 1 bar, were used. The gas velocity was 4,0 cm. sec^{-1} for both gases. The samples were flushed from the sample-holder into the pre-heated reactor and reached temperature within a minute . The reactions were stopped by removing the reactor from the oven.

For comparison, we carried out some reactions under vacuum in a rotating 50 ml flask. The flask was attached to a rotavap apparatus and heated in a silicon oil bath, the vacuum applied was ·3 mbar.

Endgroup analyses were carried out with an automatic potentiometer. The [-NH$_2$] and [-COOH] were determined simultaneously. The polymer was dissolved in o-cresol/chloroform mixture (70/30), excess alcoholic KOH added and titrated with alcoholic HCl (0.1 N). The inherent viscosities (η_{inh}) were determined in 0.5% solutions in 90% formic acid. We observed the following relationship:

$$\log \overline{M}_n = 1.122 \log \eta_{inh} + 4.182 \tag{10}$$

Table I   Reaction Data

	Temp ($^{\circ}$C)	Reaction time (hours)	Viscosimetry		Endgroup analysis		
			η inh	$\overline{M}_n$	[COOH] eq/10^5 gr	[NH$_2$] eq/10^5 gr	$(\sqrt{P})^{-1}$
Prepolymer A	–	–	0.177	2,175	43.50	58.90	1,960
	220	.83	0.377	8,200	9.78	13.56	8,700
		3.0	0.823	12,200	6.45	11.92	11,400
		6.0	0.917	13,800	4.32	10.17	15,100
		21.5	1.444	23,000	4.00	6.12	20,200
Prepolymer B	–	–	0.165	2,010	48.20	57.80	1,890
	190	1	0.374	5,000	19.36	18.49	5,300
		2	0.442	6,100	–	–	–
		4	0.438	6,000	–	–	–
		8	0.575	8,200	12.22	13.85	7,700
		24	0.728	10,600	–	–	–
	205	1	0.429	6,800	13.73	15.48	6,900
		2	0.564	8,000	–	–	–
		4	0.661	8,600	–	–	–
		6.75	0.731	10,700	11.58	6.80	11,300
	235	1	0.732	10,700	–	–	–
		2.25	0.866	13,000	–	–	–
		4	1.024	15,600	–	–	–
		8	1.218	19,000	–	–	–
	250	1	0.900	13,500	–	–	–
		4	1.408	22,300	5.19	4.09	21,700
		7	1.601	25,800	5.26	2.52	27,500
	265	1	1.220	19,000	–	–	–
		2	1.420	22,500	–	–	–
		4	1.658	26,700	–	–	–
		8	2.461	41,800	–	–	–
	280	1	1.508	24,100	–	–	–
		2	1.769	28,900	–	–	–
		4	2.717	46,800	–	–	–
					–	–	–
		8	2.728	46,000	From u.c. $\overline{M}_n = 49,000$ $\overline{M}_w = 61,000$ $\overline{M}_z = 74,000$		

With this relationship for all samples $\overline{M}_n$ was calculated
from η_{inh}. This $\overline{M}_n$ is used for evaluating the reaction
data. The ultracentrifuge (u.c.) measurements were carried out in a
Spinco model E analytical ultracentrifuge, with 0.4% solutions in
90% formic acid containing 2.3 M KCl. By means of the sedimenta-
tion-diffusion equilibrium method of Scholte (*13*) we determine
$\overline{M}_n$, $\overline{M}_w$ and $\overline{M}_z$. The buoyancy factor ($1-\overline{v}d = -0.086$) necessary
for the calculation of these molecular weights from ultracentri-
fugation data was measured by means of a PEER DMA/50 digital
density meter.
U.V. absorptions were measured on 0.5% solutions in 90% formic
acid at 290 nm.

RESULTS AND DISCUSSION

Reaction kinetics.

The results of the reaction of finely devided powdered polymer
($0.1<d<0.2$ mm) in a stream of dry nitrogen at different reaction
times and temperatures are given in table 1. On some samples
both $\overline{M}_n$ and $(\sqrt{P})^{-1}$ have been determined but as the acid and amine
endgroup concentrations do not differ much, $(\overline{M}_n) \equiv (\sqrt{P})^{-1}$. The
highest molecular weight sample, which is most susceptible
to m.w.d. broadening, is analyzed with u.c.
The evaluation of the 220°C reaction data show (fig. 2) that the
reaction does not follow third order kinetics.
Plotting log $\overline{\sqrt{M_{n}^a}} - \overline{M}_{no}^a$ versus log reaction time (fig. 3) gives a
number of isotherms. From the slopes of these isotherms the
apparent orders were calculated and are given in table II.

Table II   Apparent orders of reaction

Reaction temperatures ($^{\circ}$C)	190	205	220	235	250	265	280
Apparent order n	5.0	5.2	4.1	4.6	3.9	3.6	3.5

With decreasing reaction temperature the apparent order was found
to increase. Thus the speed of reaction does not seem to be governed
by the kinetics of the reaction.

Diffusion of the condensate.

In order to determine wether the speed of reaction is
limited by the diffusion of the condensate, the following rate
diffusion function for sperical particles (*6*, *14*) was applied
on the reaction data for several partical sizes

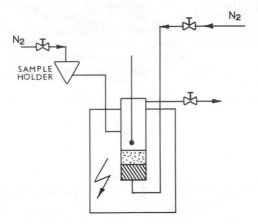

Figure 1.   Fluidized bed reactor

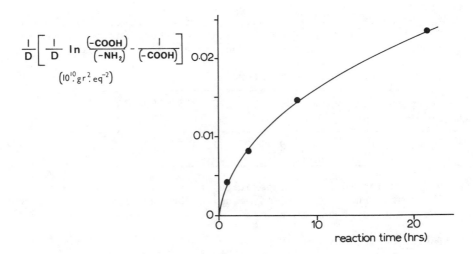

$$\frac{1}{D}\left[\frac{1}{D}\ \ln\ \frac{(-COOH)}{(-NH_2)} - \frac{1}{(-COOH)}\right]$$

$$(10^{10}.\mathrm{gr}^2.\mathrm{eq}^{-2})$$

Figure 2.   Reaction kinetics, third order relationship (reaction temperature $220°C$, $\overline{M}_{n_0}$ 2175 and particle size $0.1 < d < 0.2$ mm)

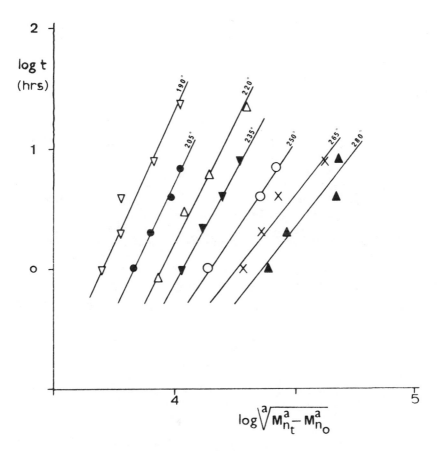

Figure 3.   Reaction kinetics (particle size 0.1 < d < 0.2 mm)

$$\frac{\overline{M}_{n_o}}{\overline{M}_{n_t}} = \frac{6}{\Pi^2} \sum_{n=1}^{\infty} \frac{1}{n^2} \exp{- n^2 \frac{\Pi^2}{r^2} D_f t.} \qquad (12)$$

For longer drying times can this be simplified to

$$\frac{\overline{M}_{n_o}}{\overline{M}_{n_t}} = \frac{6}{\Pi^2} \exp{- \frac{\Pi^2}{r^2} D_f t.} \qquad (13)$$

or

$$\ln \frac{\overline{M}_{n_o}}{\overline{M}_{n_t}} = \ln \frac{6}{\Pi^2} - \frac{\Pi^2}{r^2} D_f t. \qquad (14)$$

D_f is the diffusion constante and r the partical radius.

In figure 4 can be seen that none of the particle   size ranges
follow this diffusion function. The rates of reaction in the
3.3<d<5.0 mm particles is about half of that in the 0.1<d<0.2 mm
particles which difference is relatively small.
The diffusion of the condensate in the polymer does not seem to
be the limiting process but it does play a part in the overall
reaction rate.

Molecular weight of the starting material.

We define the starting molecular weight $\overline{M}_{n_o}$ as the molecular
weight of a material as it has just been cooled from the melt.
Samples with different $\overline{M}_{n_o}$ were prepared by varying the water
concentration when preparing the polymers. We found that
the lower the starting molecular weight of the polymer the lower
the $\sqrt[a]{\overline{M}_{n_t}^a} - \overline{M}_{n_o}^a$ after one hour reaction, while the apparent order
of reaction remained constant (table III).

Table III

Influence of starting molecular weight.

$\overline{M}_{n_o}$	1400	2010	3500
$\sqrt[a]{\overline{M}_{n_t}^a} - \overline{M}_{n_o}^a$ after 1 hour reaction at 250°C	10,000	12,700	13,800
apparent order	3.5	3.9	3.8

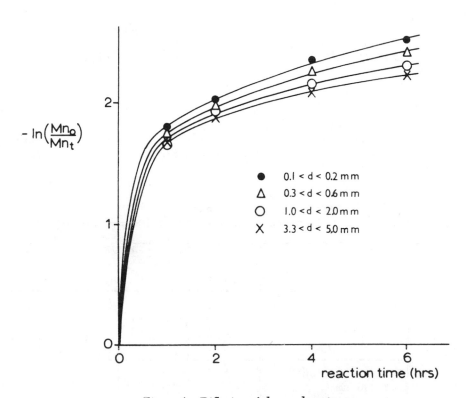

Figure 4.   Diffusion of the condensate

This effect of $\overline{M}_{n_o}$ can be explained as being due to the crystalline phase in the semi-crystalline polymer. The presence of this crystalline phase reduces the molecular mobility. The crystalline structure is not something static, but it is perfected on annealing. The longer the reaction at a high temperature, the more perfect the crystalline phase, and the more the molecular mobility is restricted. After melting this starts all over again and the lower the $\overline{M}_{n_o}$ the faster is this crystallization process.

Molecular weight broadening.

    None of the analyzed samples showed the presence of gels and the low $\overline{M}_w/\overline{M}_n$ and $\overline{M}_z/\overline{M}_w$ ratios of the highest molecular weight sample (table I) indicate that the SPP is not susceptible to m.w.d. broadening.

Type of reactor gas.

    The SPP in a fluidized bed reactor with dry nitrogen as carrier gas allows us to study the reaction under anhydrous conditions. However under these conditions the products were found to be coloured (table IV).

Table IV
Influence of the gas-sphere
Reactions at 250°C on powdered polymer ($0,1 < d < 0,2$ mm) $\overline{M}_{n_o}$ = 1400

		Fluidized bed		Rotary vacuum dryer 0.3 m.bar
		N_2 1 bar	steam 1 bar	
After 4 hrs. at 250°C	Apparent order	3.5	-	3.3
	$\overline{M}_n$	17,600	15,700	17,900
	U.V. absorption at 290 nm	.528	.023	0.521

Reacting in the rotating flask under vacuum gave a progress of reaction similar to the synthesis in the fluidized bed. Reactions under vacuum also gave coloured products.
    The SPP in a fluidized bed reactor with super heated steam as carrier gas gave somewhat lower molecular weight products but the samples showed hardly any UV absorption. Steam in the carrier gas reduces the overall polymerization rates, but suppresses the reactions which lead to coloured products (15).

Conclusions.

    The polyamidation of nylon 4,6 in the solid-phase showed that the rate of reaction is strongly dependent on the starting molecular

weight of the material, the reaction temperature and the reaction
time and to a lesser extend on the particle size and the water
concentration in the reactor gas.
Thus, the reaction rate seems not to be determined by the
of the reaction, nor by the diffusion of the condensate. This, and the
susceptibility of reaction rates to variations in starting
molecular weights make us believe that the process limiting
step is the diffusion of the reactive endgroups.
In semi-crystalline polymers at least two effects play a role
in the diffusion of the reactive endgroups. Firstly, the restriction
in endgroup movement due to the lowering of the temperature, which
usually follows an Arrhenius type equation. Secondly, the restriction
of the molecular mobility as a result of the presence of the
crystalline phase whose size and structure changes on annealing.
The endgroup diffusion restricts the number of endgroups which
takes part actively in the reaction. The rate of reaction
depends   on the active endgroup concentration which does not only
change with progress of reaction but also by the diffusion of
endgroups.
Changing the equilibrium conditions by having condensate in the
sample due to water in the carrier gas or the diffusion limitation
of the condensate in larger particles changes the reaction speed.
Although the kinetics of the reaction and the diffusion of the
condensate are not the process limitating steps they have an
effect on the overall reaction rate as described above.

ACKNOWLEDGEMENT

The authors wish to thank Dr. Ir. A.W.M. Roes, Mr. L. Bakker
and Miss Mirja Salonen for taking part in the project,
Mr. G. van de Ridder for all the molecular weight determinations
and Dr. J.W.A. van den Berg for the stimulating discussions.

ABSTRACT

The polyamidation in the solid phase is studied on the high
melting and non gelling nylon 4,6 in a fluidized bed reactor.
The rate of reaction was found to be strongly dependent on starting
molecular weight, reaction temperature and reaction time and to a
lesser extend on the particle size and the waterconcentration in
the reactor gas.
The main process limitating step seems to be the diffusion
of the reactive endgroups.

LITERATURE CITED

1 JACOBS D.B. and ZIMMERMAN J, Polymerization Processes. High
   Polymers XXIX, SCHILDKNECHT C.E. and SKEIST I. ed. Wiley
   Interscience p. 424.(1977)
2 FLORY P.J., U.S. PAT 2,173,374.

3 BRIT, PAT, 801,733 (BASF).

4 WILOTH F., U.S. PAT 3,379,696.

5 GRISKEY R.C. and LEE B.I., J. Appl. Polym. Sci. (1966) 10, 105

6 CHEN, F.G., GRISKEY R.G. and BEYER G.H., AIChEJ (1969) 15, 680

7 MONROE G.C. U.S. PAT, 3,031,433.

8 ZIMMERMAN J. Polym. Lett. (1964), 2, 955.

9 RAMSEY K.W. and DUNNILL J.H. U.S. PAT 3,240,804.

10 GAYMANS R.J. VAN UTTEREN T.E.C. VAN DEN BERG J.W.A. and
   SCHUIJER J., J. Polym. Sci. Chem. ed. (1977) 15, 537.

11 MEYER K., Angew. Makromol. Chem. (1973), 34, 165.

12 KHARITONOV V.M. FRUNZE T.M. and KORSHAK V.V. Bulletin Acad,
   Sci. USSR (1957) 1002.

13 SCHOLTE T.G., J. Polym. Sci A-2, (1968) 6, 91.

14 PERRY R.H. and CHILTON C.H. Chemical Engineers Handbook, 5th ed.
   McGraw-Hill, New York 1973.

15 PEEBLES L.H. and HUFFMAN M.W. J. Polym. Sci. A-1 (1971) 9, 1807.

RECEIVED January 29, 1979.

Conversion and Composition Profiles in Polyurethane Reaction Molding

MATTHEW TIRRELL, LY JAMES LEE, and CHRISTOPHER W. MACOSKO

Department of Chemical Engineering and Materials Science,
University of Minnesota, Minneapolis, MN 55455

Fabrication of most articles from polymeric materials has been done by melt forming of thermoplastic materials. Recently, technology has been developed for rapid in situ polymerization, to form the desired articles directly from monomeric liquids. This process has come to be known as Reaction Injection Molding (RIM).(1) To date, the major commercial RIM-processed materials are polyurethanes. The reasons for this are that polyurethane chemistry is able to provide both 1) fast, complete reaction with no side products, necessary to minimize mold cycle time and 2) a wide degree of modulus variability, through the domain-forming properties of segmented polyurethanes and the introduction of some crosslinking, necessary to give the desired mechanical properties. Development of similar desirable characteristics in epoxy, silicone, polyester, nylon (and perhaps other) polymerizations is currently a very active field (2). Principal application of this technology has been in the automotive industry. For example, polyurethane automotive facia as large as fifteen pounds are presently being reaction injection molded in a single shot in less than one minute from liquid components. Nearly fifty million pounds of polyurethane parts will be produced in the US by RIM in 1978.(3) The low temperature and pressure requirements of RIM lead to economic advantages of lower capital equipment and energy costs, relative to thermoplastic injection molding and metal forming. The lower density of polymers in general, or, more exactly, their high specific strength, provides an additional energy conservation motivation for transportation application of more polymeric materials.(2)

Vital to the continued growth of RIM processing is a basic understanding of how the process influences the structure and properties of the polymer formed. Fusion of the skills of polymerization reactor designer and injection molding designer is required. In this paper, we focus primarily on the polymerization engineering aspects although naturally the two cannot be entirely divorced. More specifically, we describe here experi-

0-8412-0506-x/79/47-104-149$07.75/0

ments and modelling efforts aimed at elucidating the effects of
nonuniform spatial distributions of temperature and conversion
during linear polyurethane RIM polymerization on the molecular
weight, molecular weight distribution sequence distribution,
and ultimately, morphology and mechanical properties of RIM
produced polyurethane.

Brief Description of RIM Process and Reactant Chemistry

A brief description of the actual physical conditions of
the RIM process is in order here. Figure 1 shows the key elements of the RIM polymerization process schematically. There
are two reactant reservoirs. One contains a diisocyanate (commercially often 4,4'-diphenylmethanediisocyanate, MDI) and the
other a mixture of polyols (one low molecular weight short diol,
and one, more flexible macrodiol). We are thus dealing with a
step-growth copolymerization. (Trifunctional polyols are also
often used.) Reactants are metered in, in the exact stoichiometric ratio necessary to achieve high molecular weight, from
each reservoir with a single stroke of the drive cylinder. An
effective bench top laboratory RIM machine, developed at the
University of Minnesota, (4) which utilizes a pneumatically driven
cylinder and a movable lever arm for stoichiometry control is
shown in Figure 2. The two reactant streams impinge on one
another in the mixing head, flow through the runner and fill
the mold in about two seconds. When the urethane polymerization
is catalyzed for example by dibutyltindilaurate, the reaction is
very fast, begins at the moment of impingement, proceeds somewhat in the runner, reaches a very high conversion in the mold
and solidifies in as little as ten seconds. The part is then
ejected from the mold and the mold resealed. This completes one
cycle from the point of view of the machine. Reaction is usually incomplete in this solid state and is driven to completion
by a post-RIM curing. Of course, the degree of conversion
which has been achieved in the time of one cycle dictates the
molecular weight of the polyurethane formed in this step growth
copolymerization.(5) Moreover, since the urethane-forming
reaction is highly exothermic and RIM-formed parts often have
substantial thickness and low thermal conductivity, a nonuniform
temperature profile develops across the part during polymerization. Therefore, at any specified reaction time, a higher conversion (and therefore higher molecular weight polymer) will be
achieved in the higher temperature regions of the polymerizing
mixture. Our modelling and experimental efforts, (to be described below), have provided a quantitative basis for these
qualitative statements.

There is however another influence on the course of urethane RIM polymerization which is perhaps not so obvious. It
results from the chemistry and structure of the reactants themselves. The structures of the reactants used in this work are

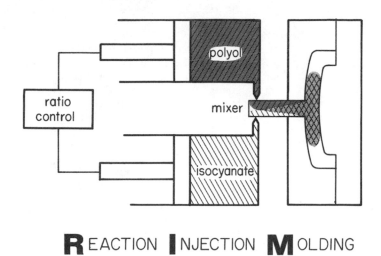

REACTION INJECTION MOLDING

Figure 1. Schematic of RIM machine

Figure 2. Photograph of University of Minnesota Laboratory RIM machine

shown in Figure 3. These are typical, as noted above, of those
used in commerical RIM formulations. They copolymerize to form
sequence length distributions of AABB units (hard segments) and
AACC units (soft segments). It has long been recognized that,
in the solid state, such a segmented copolyurethane will form a
phase separated structure, where the hard segments of sequence
length separate to form semi-crystalline domains in a less crystalline matrix of soft segment and short hard segment materi-
al(6). This type of structure is responsible for the desirable
mechanical properties of these materials. Schnieder and co-
workers(7) have proposed that the overall organization of the
solid-state of a segmented polyurethane may be spherulitic as
shown in Figure 4.

    Very little is known, on the other hand, about the "mor-
phology" of the polymerizing mixtures, or for that matter, how
the reactor conditions influence the structure and morphology
of urethane polymers. It is known from the relatively few gel
permeation chromatography studies of polyurethanes that have
been done that, while the polydispersities of MWD of polyurethanes formed in solution are very close to 2(8), characteristic of the geometric distribution, bulk polymerizations give
products with very broad, often bimodal, MWD of polydispersity
in the range of 6 to 20.(9) We here present evidence that, in
fact, phase separation occurs at a relatively early stage in the
polymerization and exerts a profound influence on the course of
the subsequent polymerization, the molecular weight distribution
(and morphology) of the polymer formed and ultimately the
mechanical properties of product formed.

Temperature Profiles During RIM Polymerization-
Analysis and Experiment for a Nonlinear Polymerization

    The heat transfer problem which must be solved in order to
calculate the temperature profiles has been posed by Lee and
Macosko(10) as a coupled unsteady state heat conduction problem
in the adjoining domains of the reaction mixture and of the
nonadiabatic, nonisothermal mold wall. Figure 5 shows the geometry of interest. The following assumptions were made: 1) no
flow in the reaction mixture (typical molds fill in ≤ 2 sec.);
2) homogeneous well mixed reaction system; 3) negligible molecular diffusion; 4) n^{th} order reaction kinetics; 5) one-dimen-
sional heat conduction (thin slab-like parts); 6) constant
thermal properties and; 7) turbulent flow of cooling fluid.
The equations have been formulated and solved numerically.
Temperature profiles throughout the system as a function of time
are shown in Figure 6. It is seen that centerline temperature
excursions to as high as $100^{\circ}C$ above the initial temperature and
differences of more than $50^{\circ}C$ between the centerline and mold
wall temperatures are possible. Note also that the mold wall
is far from isothermal.

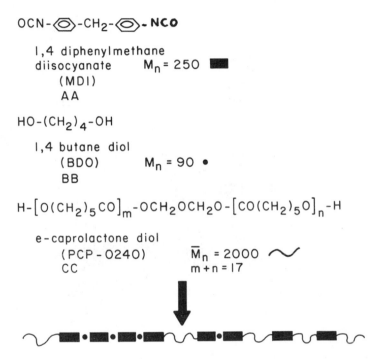

$$OCN-\langle\bigcirc\rangle-CH_2-\langle\bigcirc\rangle-NCO$$

1,4 diphenylmethane
diisocyanate $M_n = 250$ ■
    (MDI)
    A A

$$HO-(CH_2)_4-OH$$

1,4 butane diol
    (BDO) $M_n = 90$ •
    BB

$$H-\left[O(CH_2)_5CO\right]_m-OCH_2OCH_2O-\left[CO(CH_2)_5O\right]_n-H$$

e-caprolactone diol
    (PCP-0240) $\overline{M}_n = 2000$ ∼
    CC $m+n = 17$

Figure 3.   Reactants used and schematic of segmented polyurethane

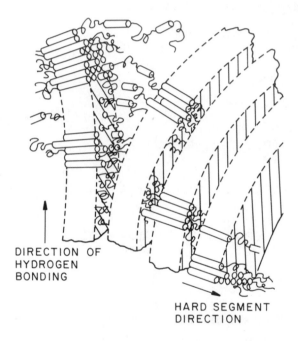

DIRECTION OF
HYDROGEN
BONDING

HARD SEGMENT
DIRECTION

Figure 4.   Morphological model for segmented polyurethanes (7)

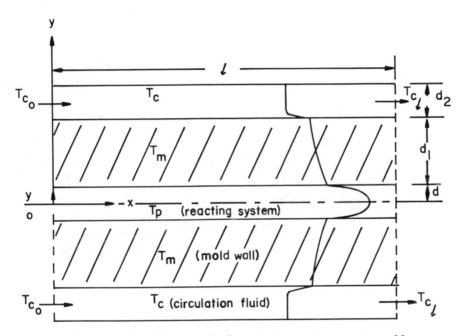

Figure 5.   Two-dimensional schematic of a reaction injection mold

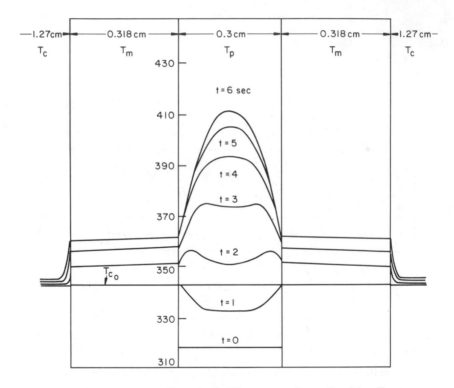

*Figure 6.   Temperature profile in the mold for nonisothermal mold wall tempera-
ture (steel mold), $V_C = 10$ cm/sec and $(T_P)_{max} = 413°K$ at 7 sec*

These results have been fit to experimental data obtained
for the reaction between a diisocyanate and a trifunctional
polyester polyol, catalyzed by dibutyltindilaurate, in our
laboratory RIM machine (Figure 2). No phase separation occurs
during this reaction. Reaction order, n, activation energy, Ea,
and the preexponential factor, A, were taken as adjustable parameters to fit adiabatic temperature rise data. Typical comparison between the experimental and numerical results are shown in
Figure 7. The fit is quite satisfactory and gives reasonable
values for the fit parameters. Figure 8 shows how fractional
conversion of diisocyanate is predicted to vary as a function of
time at the centerline and at the mold wall (remember that
molecular diffusion has been assumed to be negligible).

Obviously, quite large differences in conversion can exist
across a part, in principle. In practice, however, it is experimentally difficult to measure conversion vs time on this
time scale in a RIM mold. In a homogeneous step-growth polymerization, reaction conversion alone dictates the MWD(5). Thus,
we might expect there to be large differences in average molecular weight across the RIM part. One way to reduce temperature
gradients in the part is to heat the mold. The predicted effect
of three different circulation fluid temperatures on these conversion profiles is also illustrated in Figure 8.(10) Note that
if the mold wall is heated $25^{\circ}C$ above the entering temperature
of the reactants there is little effect on the centerline conversion but significant reduction in the difference between the
center and the wall. Thus, with some mold heating, property
development may be expected to be much more uniform. The implications of this model for a RIM polymerization with phase
separation remain to be fully explored.

This set of reactants produces a network polymer unsuitable
for solution characterization and molecular weight determination,
and without phase separation. We therefore have undertaken a
new set of experiments in RIM polymerization using the reactants
illustrated in Figure 3. As mentioned previously, the segmented
character of these products leads to phase separation in the
solid state (Figure 4) and desirable elastomeric properties in
a linear polymer. This reaction scheme represents a linear
copolycondensation of the AA, BB, CC variety where BB and CC both
react with AA (perhaps at different rates) but not with one another. Before presenting experimental data on the effects of
temperature variations during polymerizations on the structure
and properties of these polymeric products, we devote a section
to the modelling results for the molecular weight distribution
and sequence distribution for a homogeneous, isothermal polymerization of this variety. This will provide one yardstick
against which to measure the importance of effects of nonuniform
spatial distributions of temperature and concentration.

Modelling Results for Homogeneous Isothermal AA, BB, CC

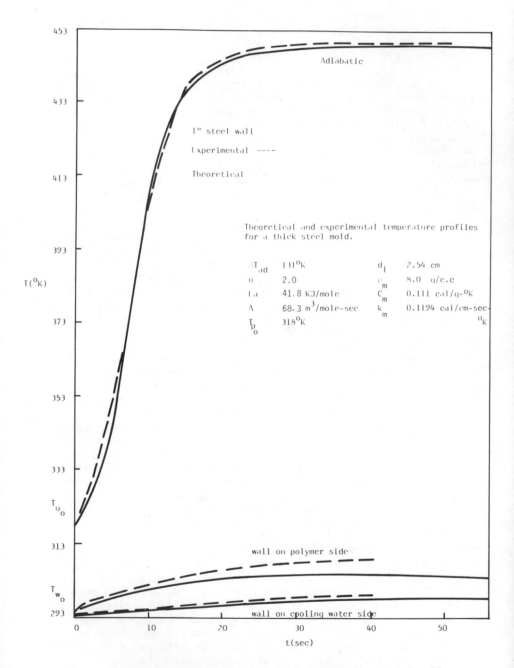

Figure 7.   *Theoretical and experimental temperature profiles for steel mold*

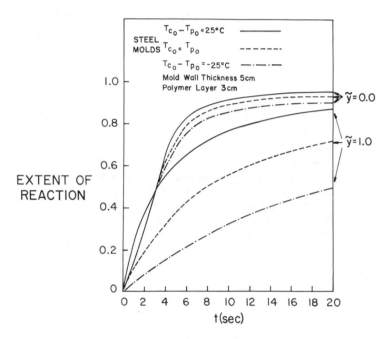

Figure 8.   Conversion vs. time (predicted) at mold center ($\bar{y} = 1.0$) and mold wall ($\bar{y} = 1.0$)

Copolymerization

The modelling of this polymerization may be approached by
deterministic or stochastic means. In previous analyses of
this reaction, Case has developed expressions for the chain
length distribution from statistical arguments (11). Peebles
(12) has formulated and solved the differential equations for
the number average sequence length, with various alternative
defining criteria for determining whether or not a unit partici-
pates in a hard segment. He used statistical arguments to show
the hard segment lengths had a geometrical length distribution.
If one is interested in the average values of the distributions
of chain length, molecular weight and sequence length, expressions for all of these can be derived quite simply by developing
recursion relations for the expected lengths of various reaction
sequences. This is an approach to polymerization modelling that
was developed by Miller and Macosko (13) for nonlinear polymerization; its application to a wide range of problems in copolymerization statistics and its relation to other stochastic modelling
methods will be fully explored in a forthcoming publication.
It is especially simple to apply to linear polycondensation.
Table I contains the results, some of which have been derived
previously (11, 12). These equations are appropriate for the
case of equal reactivity of both ends of a difunctional molecule
and allow for unequal rate constants for the A-B and A-C reactions. These results are presented here in terms of reaction
probabilities, p_{IJ}, (the probability that reactant I has reacted
with reactant J) where I,J = A, B or C, and $p_{AJ} = p_{JA}$. These
should be distinguished from the sequential probabilities of
Peller. (14) These reaction probabilities can easily be calculated
from any of the fractional conversions of A,B, and C endgroups; p_I ,
q_B and q_C respectively, if the reaction order is known. For
example, it is obvious that $p_{JA} = q_J$. From stoichiometry,

$$p_I = (\frac{1}{r_B})q_B + (\frac{1}{r_C})q_C . \qquad (1)$$

Another relation between the conversions may be developed if the
reaction order is known. For example, if the kinetics are
first order with respect to the concentrations of BB and CC
it can easily be shown that:

$$1 - q_B = (1 - q_C)^\rho , \qquad (2)$$

where $\rho = k_B/k_C$, the ratio of reaction rate constants for the
reactions of A with B and C. Given equations (1) and (2) and
the definitions given in the text above, all of the p_{IJ} can be
written in terms of a conversion of a single reactant, say that
of   isocyanate p_I.

TABLE I
Average Values of Distributions
for Homogeneous Step Growth
AA,BB,CC Copolymerization

Molecular Weights:

$$\overline{M}_w = (1 - r_B P_{AB}^2 - r_C P_{AC}^2)^{-1} \left\{ W_{AA}[M_{AA}(1 + r_B P_{AB}^2 + r_C P_{AC}^2) \right.$$
$$+ 2[M_{BB}P_{AB} + M_{CC}P_{AC}]]$$
$$+ W_{BB}[2r_B P_{AB}(M_{AA} + M_{CC}P_{AC}) + M_{BB}(1 + r_B P_{AB}^2 - r_C P_{AC}^2)]$$
$$\left. + W_{CC}[2r_C P_{AC}(M_{AA} + M_{BB}P_{AB}) + M_{CC}(1 - r_B P_{AB}^2 + r_C P_{AC}^2)]\right\}$$

(1)

$$\overline{M}_n = (1 - r_B P_{AB}^2 - r_C P_{AC}^2)^{-1} \left\{ n_{AA}[M_{AA} + M_{BB}P_{AB} + M_{CC}P_{AC}] \right.$$
$$+ n_{BB}[r_B P_{AB}(M_{AA} + M_{CC}P_{AC}) + M_{BB}(1 - r_C P_{AC}^2)]$$
$$\left. + n_{CC}[r_C P_{AC}(M_{AA} + M_{BB}P_{AB}) + M_{CC}(1 - r_B P_{AB}^2)]\right\}$$

(2)

or

$$\overline{M}_n = \frac{M_{AA} + \frac{1}{r_B} M_{BB} + \frac{1}{r_C} MCC}{1 + \frac{1}{r_B} + \frac{1}{r_C} - 2(P_{AB} + P_{AC})}$$

(3)

Chain Lengths:

Equations for $\overline{DP}_n$ and $\overline{DP}_w$ are identical to the corresponding ones from (1), (2) and (3) above with all the M_{AA}, M_{BB}, M_{CC} replaced by a factor of unity.

Sequence Lengths: (AABB Sequences)

$$\overline{N}_w = \frac{1 + r_B P_{AB}^2 + 2P_{AB}}{1 - r_B P_{AB}^2}$$

(4)

$$\overline{N}_n = \frac{1}{1 - r_B P_{AB}^2}$$

(5)

Note: $r_B = [AA]_o/[BB]_o$; $r_C = [AA]_o/[CC]_o$; W_{ii} and N_{ii} are the weight and mole fractions of species ii, respectively.

Several features of these equations bear mention. At high conversions, the ratios of the weight average to number average values of all three distributions go to 2, indicating that the chain length and sequence length distributions are always nearly geometrically (or "most probably") distributed. This result has been discussed previously by Peebles (12) for the sequence length distribution. These equations are much simpler to use than his deterministic treatment and give the same simple limiting results. Under conditions where reaction temperature varies spatially, these equations predict that molecular weight will also vary spatially (with negligible molecular diffusion) for two reasons: 1) conversion will vary spatially at any specified reaction time and 2) the parameter ρ will have some temperature dependence, so that even if spatially uniform final conversion is eventually achieved, different temperature histories in different areas of the reaction medium can be lead to a part with nonuniform molecular weight and sequence length distributions, due to the different relative rates for adding BB and CC into the chain. These are the most important results of the homogeneous model of urethane polymerization.

There is very little experimental data available on values of ρ for these reactants. Some isothermal data indicates that values in the neighborhood of 3 to 4 are reasonable (15), but virtually nothing is reported in the literature on the temperature dependence. This makes quantitative comparison with data more difficult, however certain aspects such as the polydispersity prediction of 2 are easily checked. Thus, we now will examine the utility of this model under various experimental polymerization conditions.

Experimental Results on the Effects of Spatial Temperature Variations on RIM Polymerization

In order to get a quantitative idea of the magnitude of the effects of these temperature variations on molecular structure and morphology an experimental study was undertaken. Two types of polymerizations were conducted. One type was isothermal polymerization at fixed reaction time at a series of temperatures. The other type was a nonisothermal polymerization in the geometry of a RIM mold. Intrinsic viscosities, size exclusion chromotograms (gpc) and differential scanning calorimetry traces (dsc) were obtained for the various isothermal products and from spatially different sections of the nonisothermal products. Complete experimental details are given below.

Polymerization. Isothermal polymerizations were conducted in 1/8" molds controlled at the stated isothermal temperature. Below 100°C circulating water was the temperature control medium whereas circulating air was used above 100°C. Reactants were

mixed and degassed at $65°$ before entering the mold,
thus some small degree of reaction takes place outside the mold.
Isothermal reaction times were 20 hours in all cases. To slow
down the polymerization to a rate enabling us to follow time-
dependent changes more easily, no catalyst was used in these reactions. Nonisothermal reactions were conducted in a 1/2" thick
mold with walls of 5/16" aluminum. Centerline and mold wall
temperatures were recorded by thermocouples as functions of
time. Two different mold wall temperatures were used: $85°C$
(run number N-1) and $37°C$ (run number N-2). The reactants
of Figure 3 were used in all cases. Two sets of CC/BB/AA
initial mole ratios were used 1/5/6 and 1/2/3 providing different relative amounts of hard and soft segments. Isothermal
runs are designated by a number giving the temperature in $°C$
and the initial mole ratio (e.g. 70-156). All nonisothermal
runs were done with the 1/5/6 proportions.

GPC: A Dupont 830 HTSEC Liquid Chromatograph was used
for the gpc analyses. Stationary phase was four Dupont silica
25cm. SEC columns in series with nominal pore sizes of 1000Å,
500Å, 100Å and 60Å. Mobil phase was tetrahydrofuran (THF)
containing 1% by weight of (Polysciences) polyethylene oxide
1000 to minimize adsorption. Elution was at 1ml/min. at $30°C$.
Calibration was done with polystyrene standards and thus the
molecular weight values obtained directly were polystyrene
equivalent molecular weights. These were converted to polyurethane molecular weights (the numbers reported here) by
eluting a sample of polyurethane of molecular weight determined by intrinsic viscosity (see Eq. 3 below). The correction factor multiplying the "PS equivalent" molecular weights
used was 0.23. The solubility of some of these polyurethanes
in THF is rather limited (9). All solutions for gpc were
prepared by placing each polyurethane sample in contact with
enough THF to give a 0.1 wt.% solution (if fully dissolved)
and allowing to dissolve for at least 24 hours at $60°C$ in sealed
vials. Not all samples completely dissolved under these conditions. In cases where dissolution was incomplete, the solution
was filtered and a sample of the clear supernatant analyzed
by gpc. This accounts for the relatively small area under some
of the gpc traces. All curves shown are at the same sensitivity scale. This limited solubility is surely a reflection of
the morphology and thus, probably indirectly of the sequence
distribution and/or molecular weight but we are unable to
give a detailed interpretation at present.

Intrinsic Viscosity: Measurements were made by standard
techniques in dimethyl formamide (DMF) at $30°C$. Weight
average molecular weights were calculated from the following
relationship:

$$[\eta] = 6.80 \times 10^{-5} \bar{M}_W^{0.86} \qquad (3)$$

This equation is based on a light scattering-intrinsic viscosity correlation for pentanediol based polyurethanes (9). No
molecular weight degradation in DMF was observed after aging
the solutions for several days.

DSC: Calorimetry was done in a Perkin-Elmer DSC-II at a
scan rate of 20°C/min. The scale used is 2mcal/sec. full
scale. Sample weight was 12~20 mg. with an empty sample pan
as reference.

Results: All the numerical results on molecular weights
and molecular weight distribution by intrinsic viscosity and
gpc are given in Table 2. GPC traces of the samples are
given in Figures 9 & 10. There is acceptable agreement between the gpc and $[\eta]$ molecular weights considering the difficulties inherent in both methods for obtaining absolute values
for molecular weights of copolymers of different compositions.
Of course, this does not affect the comparison between samples
by either technique. For the isothermal samples of Figure 9,
several trends are apparent. Most obviously, the average
molecular weights for samples of equivalent stoichiometry
are higher with increasing reaction temperature. This
may be at least partly due to increased conversion at higher
temperature. The resolution of the gpc traces at the low
molecular weight end is good enough to discern what appear to
be lower molecular weight oligomers of several discrete molecular weights. They decrease in concentration as molecular
weight (and conversion) increases. It is obvious that the
samples with higher hard segment content have 1) broader MWD
and 2) more temperature sensitive molecular weights. This
is evidence of influence of the polymerization medium morphology on the course of the reaction.

The temperature profiles for the nonisothermal samples
N-1 and N-2 are shown in Figure 11. It is seen that the mold
wall remains isothermal and that some exothermic reaction has
begun before the reactants enter the mold. The centerline
temperature reaches 158°C with N-2 and 170°C with N-1. In
the nonisothermal samples of Figure 10, it is seen that, in
fact, very large differences in molecular weight, roughly a
factor of two, can exist between the surface and center of
sample N-2 (cold mold walls) and that much of this difference
can be eliminated by heating the mold wall, as demonstrated
with sample N-1. This is consistent with our speculation
based on Figure 8. Note that the limited solubility of several
samples has reduced the area under the gpc trace. For this
reason, the molecular weights determined by intrinsic viscosity
are probably more representative of the molecular weight of the
entire sample.

An observation with significant implications is that the
dispersity of the MWD of all the samples is considerably
greater than 2, except for the 1/2/3 samples polymerized at

TABLE 2

Molecular Weight Results on Polyurethanes

From Different Reaction Conditions

Sample Designation	Intrinsic Viscosity (DMF)		GPC (THF)			Comments
	$[\eta]$ (dl/g)	$\bar{M}_w$	$\bar{M}_n$	$\bar{M}_w$	Polydispersity	
70-156	0.201	10,850	2,150	5,200	2.40	—
90-156	0.311	18,000	5,600	23,900	4.27	—
110-156	0.396	23,890	4,600	11,040	2.40	*
70-123	—	—	6,440	14,200	2.20	—
85-123	—	—	7,380	20,860	2.83	—
110-123	—	—	6,200	12,470	2.01	*
130-123	—	—	7,840	14,770	1.90	*
N1 center	0.359	21,350	4,260	9,360	2.20	**
N1 surface	0.362	21,500	3,540	9,000	2.53	*
N2 center	0.477	27,500	12,850	48,700	3.80	**
N2 surface	0.273	15,500	4,140	13,660	3.30	—

(Note: all gpc results are on THF soluble fraction only)

---

* some sediment in THF solution
** much insoluble in THF

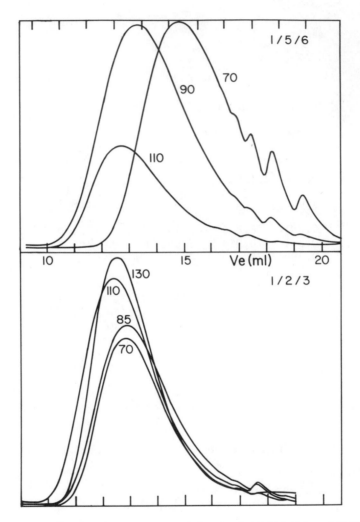

Figure 9.   GPC results on isothermal samples: (top) CC/BB/AA = 1/5/6; (bottom) CC/BB/AA — 1/2/3/. *Numbers by curves indicate polymerization temperature in degrees C.*

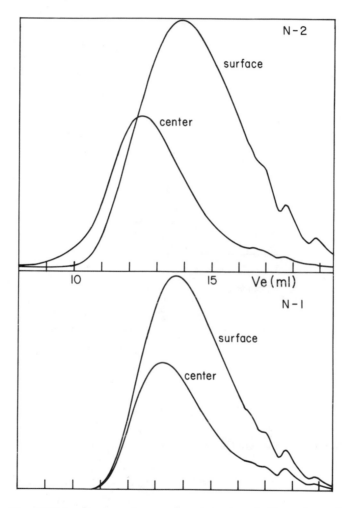

Figure 10. GPC results on surfaces and centers of nonisothermal samples (both are 1/5/6)

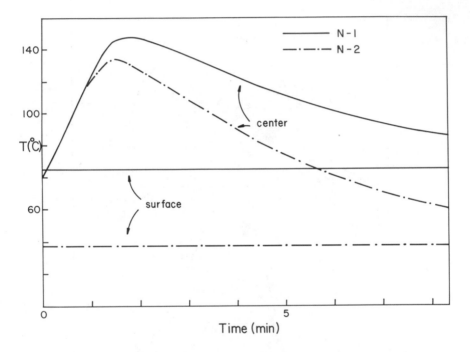

Figure 11.   Measured temperature profiles for samples N-1 and N-2

high temperatures. Although the variation in dispersity among
the samples is not very great, there does seem to be a clear
trend in the lower hard segment content 1/2/3 samples for the
MWD to narrow as temperature increases. This may reflect
changes in the morphology of the polymerizing mixture with
temperature, as we shall see in the dsc results to be presented.
These high values of polydispersity of MWD are the first indication of the inadequacy of the homogeneous polymerization
model of Table 1.

Few direct means exist for evaluating the sequence distributions of this sort of copolymer. No doubt, however, spatial
differences in the average sequence lengths also exist under
these nonisothermal conditions. Differential scanning calorimetry was undertaken to evaluate morphology differences due to
temperature variations during polymerization and hopefully
evaluate some of the implications for the sequence length
distribution. The dsc results for several of the samples of
Table 2 are shown in Figures 12 and 13. In looking at all
of the samples, four distinct transition temperatures are observed, consistent with the observations of others on segmented polyurethanes (6). A "soft segment" glass transition
at about -58°C (not shown), a "soft segment" melting at about
45-50°C, a "hard segment" glass transition at about 90-107°C
and a "hard segment" melting at about 200-215°C, are clearly
discernable. We note that some of these transitions in the
solid polymer occur in the same temperature range that is used
for polymerization. It is clear that, in the as polymerized
material, there is much greater hard segment crystalline organization, and much poorer soft segment organization, in the
samples which are polymerized at higher temperature. This is
true both in comparing the two isothermal samples (Figure 12)
as well as in comparing the centerline and wall samples
from run number N-2 (Figure 13). A detailed interpretation of
the dsc traces will not be given here. The following observations of some significance can, however, be made at this point.
1) There is evidence, from the pre-melting exotherms observed,
that recrystallization of both the "soft segments" and the
"hard segments" is occuring during the dsc scan, especially
in the N-2 wall and 70-156 samples. This points out the
fact that the as-molded morphology of a RIM article is not
immutable but rather is subject to annealing and other thermal
treatments. This has been confirmed in our laboratories by
doing multiple dsc scans of the same sample. Differences are
observed from scan to scan, especially initially. This point
is being pursued in more detail currently. 2) There seem to
be multiple "hard segment" melting peaks, consistent with reports of other workers. At this point, however, we must also
allow for the possibility of the exotherm representing further
reaction of previously unconverted material. This is an alternative which has been given little attention previously, but is

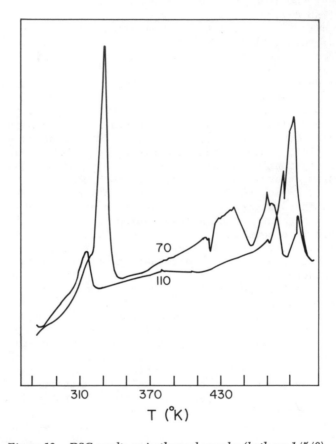

Figure 12.   DSC results on isothermal samples (both are 1/5/6)

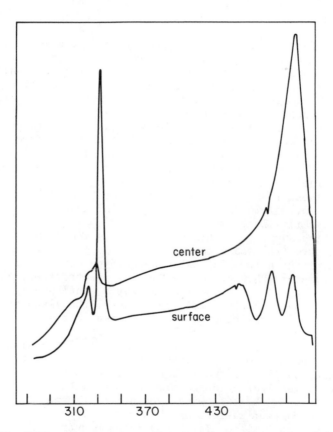

Figure 13.   DSC results on surface and center of nonisothermal sample N-2 (cold mold wall, 1/5/6)

consistent with our observation in Figures 11 and 12 of much
larger exotherms in samples polymerized at lower temperature.

In any case, it is clear from this work that spatial variations in molecular weight and morphology do exist in RIM molded
parts. Of course, both of these exert some degree of control
over the mechanical properties, which must then also be considered to be nonuniform.

Discussion

The molecular weight and morphological variations are in
themselves interesting. Seeking a quantitative explanation
for them, and for the observed MWD effects, is quite challenging however. We have noted, first of all, that in all cases,
not just the nonisothermal samples, the MWD is broader than
can be predicted by the model developed in a previous section.
We also see in general narrower MWD's when a) the temperature
is increased for the 1/2/3 samples and b) when the hard segment
content is reduced (1/5/6 to 1/2/3) at the same reaction temperature. The sensitivity of molecular weight to temperature
is much larger in the 1/5/6 samples with higher hard segment
content. These last points begin to suggest that the morphology
of the nascent polymer is exerting some influence on the course
of the polymerization.

Striking support of this contention is found in recent
data of Castro (16) shown in Figure 14. In this experiment,
the polymerization (60-156) has been carried out in a cone-and-
plate viscometer (Rheometrics Mechanical Spectrometer) and
viscosity of the reaction medium monitored continuously as a
function of reaction time. As can be seen, the viscosity appears to become infinite at a reaction time corresponding to
about 60% conversion. This suggests network formation, but
the chemistry precludes non-linear polymerization. Also observed in the same conversion range is very striking transition of the reaction medium from clear to opaque.

Given this clear evidence of inhomogeneity development,
we have developed the following hypothetical scenario for
urethane RIM polymerizations of this sort. The polymerization
proceeds in a compositionally homogeneous manner, but generating heat and possibly temperature not uniformities. At some
critical value of conversion, which may be temperature dependent,
probably dictated by some quality of the hard segment sequence
length distribution, a separation into 2 phases occurs. One
phase is composed predominantly of an organized array of hard
segments perhaps resembling Figure 4. Amorphous and crystalline
soft segments and short sequences of hard segments form a
second phase. But polymerization proceeds, now at two different rates of conversion in the two different phases, due to
the different affinities and mobilities of each reactant in
each phase. Local reaction temperature influences the degree

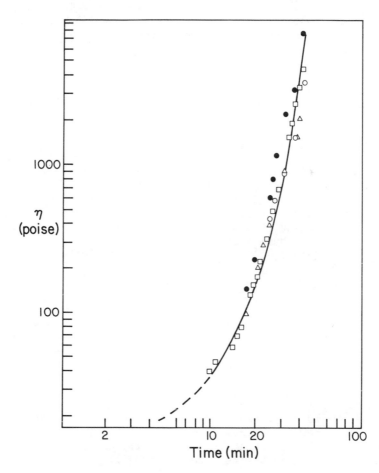

Figure 14.   Viscosity vs. reaction time for 1/5/6 polymerization at 60°C. Conversion at point of sharp viscosity rise approximately 0.60.

of organization and molecular mobility in each phase. This in
turn influences the local molecular weight and sequence distributions formed. The sharp viscosity rise must be due to some
type of interchain connectivity of the reaction medium, related to the phase separation phenomenon. The net result of
this polymerization is a broadened molecular weight distribution and an altered sequence length distribution. This is a
possible route to the very broad and bimodal MWD's observed by
other workers, with other reactants. We note in this connection
that systems with lower hard segment content have narrower MWD
and that polymerization above the hard segment Tg leads to a
product of still narrower MWD. We have never directly observed bimodal MWD but significant portions of the higher polymerization temperature samples were insoluble. There is good
reason to believe that we would have observed much broader
MWD had total dissolution been possible.

     We have put this model into mathematical form. Although
we have yet no quantitative predictions, a very general model
has been formulated and is described in more detail in Appendix A. We have learned and applied here some lessons from
Kilkson's work (17) on interfacial polycondensation although
our problem is considerably more difficult, since phase separation occurs during the polymerization at some critical value
of a sequence distribution parameter, and not at the start of
the reaction. Quantitative results will be presented in a
forthcoming publication.

Conclusions

     1. Due to the exothermic nature of the reaction and the
phase separation which occurs, temperature and conversion (and
MWD and sequence distribution) can only be assigned local, not
global, values in polyurethane reaction molding.
     2. There are two size scales for the spatial variation
of conversion in a RIM mold. One, on the scale of mold dimension, is set by the heat generation-heat removal balance and
has been shown to cause factor of two variations in molecular
weight from center to surface of a RIM part. This can be rec-
tified in practice by heating the mold wall, as shown here, or
by a post-cure reaction. The second size scale for variation
of conversion is set by the size scale of phase separation,
shown to occur during polymerization. This leads to broader
MWD and, as yet, poorly understood effects on the sequence distribution and morphology.
     3. Morphological characteristics vary with polymerization
temperature, and can vary significantly across a RIM part.
Higher polymerization temperature promotes hard segment organization in the as-polymerized material. The origin of this in
the polymerization reaction is presently unknown but under investigation.

4. A mathematical model for this polymerization reaction based on homogeneous, isothermal reaction is inadequate to predict all of these effects, particularly the breadth of the MWD. For this reason a model taking explicit account of the phase separation has been formulated and is currently under investigation.

5. One must now realize that extreme caution should be exercized in drawing conclusions about RIM polymerization from batch isothermal reactions at high temperature.

Appendix 1:   Toward a Model for Step-Growth Copolymerization with Phase Separation

A detailed description of AA, BB, CC step-growth copolymerization with phase separation is an involved task. Generally, the system we are attempting to model is a polymerization which proceeds homogeneously until some critical point when phase separation occurs into what we will call hard and soft domains. Each chemical species present is assumed to distribute itself between the two phases at the instant of phase separation as dictated by equilibrium thermodynamics. The polymerization proceeds now in the separate domains, perhaps at differen- rates. The monomers continue to distribute themselves between the phases, according to thermodynamic dictates, insofar as the time scales of diffusion and reaction will allow. Newly-formed polymer goes to one or the other phase, also dictated by the thermodynamic preference of its built-in chain micro — architecture.

Obviously, construction of a mathematical model of this process, with our present limited knowledge about some of the critical details of the process, requires good insight and many qualitative judgments to pose a solvable mathematical problem with some claim to realism. For example: what dictates the point of phase separation?; does equilibrium or rate of diffusion govern the monomer partitioning between phase?; if it is the former, what are the partition coefficients for each monomer?; which polymeric species go to each phase?; and so on.

We have begun to explore the implications of a very simple set of assumptions used to answer the above questions. In the early homogeneous stages of the polymerization, the conversion vs. time behavior is governed by the set of equations in Table A-1 (minus the I_J terms in Equations 1, 2 and 3) where the $P'_{I,J}$ are the dimensionless total concentrations of polymers with endgroups IJ. Given the conversion, average values of MWD and sequence distribution as a function of time can be predicted for this homogeneous period from the equations given previously in Table 1. We assume that phase separation occurs at some critical value of $\overline{N}_n$, say $\overline{N}^* = 4$. This gives the conversions (from Equation 5 of Table 1) when phase separation occurs. Note that the choice of $\overline{N}^*$ is independent of choosing

TABLE A-I
Equations for Step-Growth
Copolymerization with Phase
Separation

$$\frac{dp}{d\tau} = (1 - p) \left\{ 2[r_B(1 - q_1) + P'_{BB}] + P'_{AB} + P'_{BC}] \right.$$
$$\left. + K[2(r_C(1 - q_2) + P'_{CC}) + P'_{AC} + P'_{BC}] \right\} \; (\pm I_A)$$

$$\frac{dq_1}{d\tau} = (1 - q_1) \left\{ 2[1 - p + P'_{AA}] + P'_{AB} + P'_{AC} \right\} \; (\pm I_B)$$

$$\frac{dq_2}{d\tau} = K(1 - q_2) \left\{ 2(1 - p + P'_{AA}) + P'_{AB} + P'_{AC} \right\} \; (\pm I_C)$$

$$\frac{dP'_{AA}}{d\tau} = -P'_{AA} \left\{ 2[r_B(1 - q_1) + P'_{BB}] + P'_{BC} \right.$$
$$\left. + K[2[r_C(1 - q_2) + P'_{CC}] + P'_{BC}] \right\} + (1 - p) \left\{ P'_{AB} + KP'_{AC} \right\}$$

$$\frac{dP'_{BB}}{d\tau} = -P'_{BB} \left\{ 2[1 - p + P'_{AA}] + P'_{AC} \right\} + P'_{AB} \left\{ r_B(1 - q_1) + \frac{1}{2}K \; P'_{BC} \right\}$$

$$\frac{dP'_{CC}}{d\tau} = -K \; P'_{CC} \left\{ 2[1 - p + P'_{AA}] + P'_{AB} \right\} + P'_{AC} \left\{ K \; r_C(1 - q_2) + \frac{1}{2} \; P'_{BC} \right\}$$

$$\frac{dP'_{AB}}{d\tau} = -P'_{AB} \left\{ 1 - p + r_B(1 - q_1) + P'_{AA} + P'_{AB} + P'_{BB} + \frac{1}{2}(P'_{AC} + P'_{BC}) \right.$$
$$\left. + K[r_C(1 - q_2) + P'_{CC} + \frac{1}{2} \; P'_{BC}] \right\}$$
$$+ \left\{ 2[r_B(1 - q_1) + P'_{BB}] + K \; P'_{BC} \right\} (1 - p + P'_{AA})$$

$$\frac{dP'_{AC}}{d\tau} = -P'_{AC} \left\{ r_B(1 - q_1) + P'_{BB} + \frac{1}{2} \; P'_{BC} + K[1 - p + P'_{AA} + r_C(1 - q_2) \right.$$
$$\left. + P'_{CC} + P'_{AC} + \frac{1}{2}(P'_{AB} + P'_{BC})] \right\}$$
$$+ \left\{ 2K[r_C(1 - q_2) + P'_{CC}] + P'_{BC} \right\} (1 - p + P'_{AA})$$

$$\frac{dP'_{BC}}{d\tau} = -P'_{BC} \left\{ [1 - p + P'_{AA}][1 + K] + \frac{1}{2}[P'_{AC} + K\ P'_{AB}] \right\}$$

$$+ P'_{AC}[r_B(1 - q_1) + P'_{BB}] + K\ P'_{AB}[r_C(1 - q_2) + P'_{CC}]$$

Note: $K = k_2/k_1$ $d\tau = k_1[AA]_0 dt$ I_J = rate of interphase
transport of species
J.

In this Table p, q_1, q_2 are fractional conversions of AA, BB
and CC <u>monomers</u>.

$r_B = [BB]_0/[AA]_0$, $r_C = [CC]_0/[AA]_0$.

$P_I = 1 - A_T/A_0$, $q_B = 1 - B_T/\ B$.

$A_T/A_0 = 1 - p + P'_{AA} + \frac{1}{2}(P'_{AB} + P'_{AC})$;

$B_T/A_0 = r_B(1 - q_1) + P'_{BB} + \frac{1}{2}(P'_{AB} + P'_{BC})$

$C_T/A_0 = r_C(1 - q_2) + P'_{CC} + \frac{1}{2}(P'_{AC} + P'_{BC})$

$A_T = B_T + C_T$

(i.e. overall stoichiometric balance of hydroxyl and isocynate
groups).

which species go to each phase when separation occurs. We assume that any polymeric species containing an AABB sequence greater than some critical value N* will go to the hard domains, all else joins the soft domains. (Note N* may or may not = $\bar{N}_n$*). The volume fraction X of hard phase may thus be calculated from the knowledge that the sequence distribution formed in homogeneous polymerization is geometric (12).

After phase separation, two sets of equations such as those in Table A-1 describe the polymerization but now the interphase transport terms I_T must be included which couples the two sets of equations. We assume that an equilibrium partitioning of the monomers is always maintained. Under these conditions, it is possible, following some work of Kilkson (17) on a simpler interfacial nylon polymerization, to express the transfer rates I in terms of the monomer partition coefficients, and the volume fraction X. We assume that no interphase transport of any polymer occurs. Thus, from this coupled set of eighteen equations, we can compute the overall conversions in each phase vs. time. We can then go back to the statistical derived equations in Table 1 and predict the average values of the distribution. The overall average values are the sums of those in each phase.

Our preliminary results with this model indicate that distinctly bimodal MWD's are formed for some values of the parameters whereas near equality of the average values for each phase leads to a somewhat broadened unimodal MWD for other parameter choices. These results will be presented in detail, and we will explore some refinements to the above described model in a forthcoming publication (18).

Acknowledgement

The authors gratefully acknowledge the contributions made to this work by Ms. Sue Tanger and by Messrs. Franciso Lopez-Serrano, José Castro and Ron Miller and the support they received during the course of this work from the National Science Foundation (ENG77-05555 to MT, DMR 75-04508 to CWM), the Graduate School and Computer Center of the University of Minnesota (MT) and the Union Carbide Corporation (CWM).

Literature Cited

1. Wood, A.S., Modern Plastics, (1976), 53, 35.
2. MacKnight, W.J., Baer, E. and Nelson, R.D., eds., "Proceedings of a DOE Workshop Recommending Future Directions in Energy-Related Polymer Research," (1978), Document no. CONF-780643.
3. Leis, D.G., presented at Soc. Automotive Eng. meeting,

September, (1977).
4. Macosko, C.W. and Lee, L.J., (1978), U.S. patent pending.
5. Hicks, J., Mohan, A. and Ray, W.H., Can. J. Chem. Eng., (1969), 47, 590.
6. Harrell, L.J., Macromol., (1969), 2, 607; Huh, D.S., and Cooper, S.L., Polym. Eng. Sci., (1971), 11, 369.
7. Schneider, N.S., Desper, C.R., Illinger, J.L., King, A.O. and Barr, D., J. Macromol. Sci. Phys., (1975), B11, 527.
8. Vakhtina, I.A., Bettger, T., Andreyev, A.P., Novozhilova, O.S. and Tarakanov, O.G., Vysokomol soyed., (1976), A18, 2138.
9. Schollenberger, C.S. and Dinbergs, K., J. Elastoplastics, (1973), 5, 222; Seefried, C.G., Koleske, J.V., Critchfield, F.E. and Pfaffenberger, C.R., manuscript in press, (1978).
10. Lee, L.J. and Macosko, C.W., Soc. Plast. Eng. ANTEC Papers, (1978), 24, 155.
11. Case, L.C., J. Polym. Sci., (1958), 29, 455.
12. Peebles, L.H., Macromol., (1974), 7, 872.
13. Macosko, C.W. and Miller, D.R., Macromol., (1976), 9, 199.
14. Peller, L., J. Chem. Phys, (1962), 36, 2976.
15. Lenz, R.W., "Organic Chemistry of Synthetic High Polymers," Interscience, New York, 1967.
16. Castro, J.M., Ph.D. Dossier, University of Minnesota, (1978).
17. Kilkson, H., I.E.C. Fund., (1968), 7, 355.
18. Lopez-Serrano, F., Tirrell, M. and Macosko, C.W., in preparation, (1978).

RECEIVED February 6, 1979.

Phase Equilibrium in Polymer Manufacture

DAVID C. BONNER

Shell Development Company, P. O. Box 1380, Houston, TX 77001

The manufacture of synthetic high polymers involves, in most instances, combined reaction kinetics, heat transfer, and phase equilibrium. Due to the shortcomings of phase equilibrium computational methods, it has only been possible in the last decade for design engineers to apply rigorous phase equilibrium computations in the design of polymerization processes. In this paper, we discuss some of the computational methods which are now available for use in process design. It is our hope that this sort of presentation may serve to draw together some of the phase equilibrium work that has been published in the last decade in various journals.

In order to begin this presentation in a logical manner, we review in the next few paragraphs some of the general features of polymer solution phase equilibrium thermodynamics. Figure 1 shows perhaps the simplest liquid/liquid phase equilibrium situation which can occur in a solvent(1)/polymer(2) phase equilibrium. In Figure 1, we have assumed for simplicity that the polymer involved is monodisperse. We will discuss later the consequences of polymer polydispersity.

Conditions of phase equilibrium require that the chemical potential of polymer in each phase and that of solvent in each phase be equal:

$$\mu_1^\alpha = \mu_1^\beta$$
$$\mu_2^\alpha = \mu_2^\beta \tag{1}$$

The computational problem of polymer phase equilibrium is to provide an adequate representation of the chemical potentials of each component in solution as a function of temperature, pressure, and composition.

A feature of polymer solutions which is commonly observed in many polymer manufacturing operations is illustrated in Figure 2. At a given pressure, a two-phase region exists below a concave-downward locus of temperature-composition points. One of the two

0-8412-0506-x/79/47-104-181$05.00/0

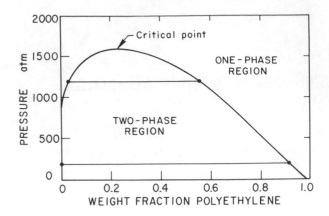

Figure 1.   Liquid–liquid equilibria in a polymer/solvent solution: ethylene–polyethylene binary coexistence curve (constant T and molecular weight)

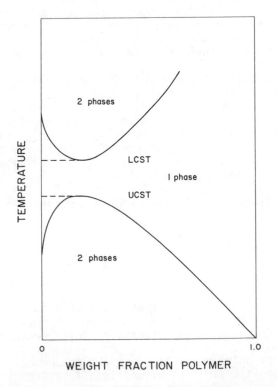

Figure 2.   Polymer/solvent phase equilibrium diagram

phases is polymer rich, and the other is solvent rich. As the temperature of a solution of constant composition is raised, a one-phase region is reached. Finally, as temperature is increased yet more, phase separation occurs again. The higher temperature two-phase region is bounded by a concave-upward locus of composition-temperature points.

The critical point (1) of the two-phase region encountered at reduced temperatures is called an upper critical solution temperature (UCST), and that of the two-phase region found at elevated temperatures is called, perversely, a lower critical solution temperature (LCST). Figure 2 is drawn assuming that the polymer in solution is monodisperse. However, if the polymer in solution is polydisperse, generally similar, but more vaguely defined, regions of phase separation occur. These are known as "cloud-point" curves. The term "cloud point" results from the visual observation of phase separation - a cloudiness in the mixture.

While the shapes of the upper and lower critical loci are most usually as shown schematically in Figure 2, a variety of other behaviors has been observed in special cases (2).

Another general type of behavior that occurs in polymer manufacture is shown in Figure 3. In many polymer processing operations, it is necessary to remove one or more solvents from the concentrated polymer at moderately low pressures. In such an instance, the phase equilibrium computation can be carried out if the chemical potential of the solvent in the polymer phase can be computed. Conditions of phase equilibrium require that the chemical potential of the solvent in the vapor phase be equal to that of the solvent in the liquid (polymer) phase. Note that the polymer is essentially involatile and is not present in the vapor phase.

Using standard thermodynamics, it can be shown (3) that, at modest pressures, the equality of solvent chemical potential in both liquid and vapor phases can be transformed to

$$p_1 = a_1 p_1^S \exp\left[B_{11}(p_1^S - p_1)/(RT)\right] \qquad (2)$$

where p_1 = solvent partial pressure

$a_1 = a_1(T, w_1, p)$ = solvent activity

p_1^S = solvent saturation vapor pressure at solution temperature T

B_{11} = second virial coefficient of solvent at T

R = gas constant

Since the total pressure ($p = p_1$) of the devolatilization process is usually known, computation of weight fraction (w_1) of solvent remaining in the polymer at the limit of phase equilib-

Solvent (1) / Polymer (2)
$$w_1 \approx 0$$

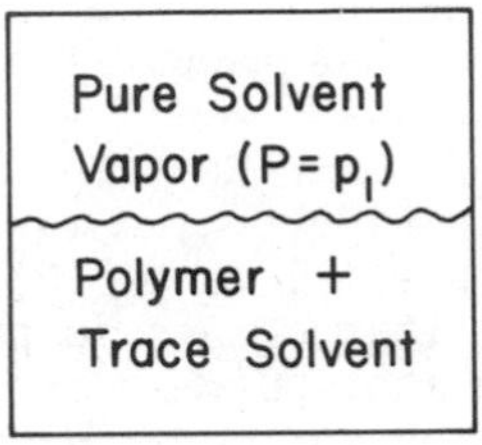

Pure Solvent
Vapor $(P = p_1)$

Polymer +
Trace Solvent

Note : polymer is
essentially
involatile

Question : Given T, P, what is the solvent
composition in polymer phase ?

Calculations : $p_1 = a_1 p_1^s \exp\left[\dfrac{B_{11}(p_1^s - p_1)}{RT} \right]$

where $a_1 = a_1(T, w_1, P) =$ solvent <u>activity</u>

$p_1^s =$ solvent vapor pressure at T

$B_{11} =$ solvent 2nd virial coefficient
at T

$P_1 = y_1 P =$ solvent partial pressure

Figure 3.   Polymer/solvent vapor–liquid equilibrium

rium requires formulation of a mathematical expression for computation of solvent activity in the polymer solution. The solvent activity is a function of temperature, pressure, and composition.

Polymer Solution Phase Equilibrium Computations.

Flory-Huggins Model for Polymer Solutions.

We have seen above in two instances, those of liquid-liquid phase separation and polymer devolatilization that computation of the phase equilibria involved is essentially a problem of mathematical formulation of the chemical potential (or activity) of each component in the solution.

The first qualitatively correct attempt to model the relevant chemical potentials in a polymer solution was made independently by Huggins (4,5) and Flory (6). Their models, which are similar except for nomenclature, are now usually called the Flory-Huggins model (2).

The Flory-Huggins activity expression for solvent in a solvent(1)/polymer(2) solution is

$$a_1 = \exp(\mu_1 - \mu_1^0/RT) = \Psi_1 \exp\left[(1 - r_1/r_2)\Psi_2 + \chi\Psi_2^2\right] \quad (3)$$

In equation (3), the term μ_1^0 is the reference chemical potential. We take the reference state for which the reference chemical potential is computed to be pure, saturated solvent at the temperature of the solution.

The segment fraction of solvent (Ψ_1) in solution is given by

$$\Psi_1 = \frac{r_1 x_1}{r_1 x_1 + r_2 x_2} \quad (4)$$

where r_1 is the number of repeating segments per solvent molecule, r_2 is the number of repeating segments per polymer molecule. The mole fractions of solvent and polymer are x_1 and x_2, respectively.

According to Flory-Huggins theory, the heat of mixing of solvent and polymer is proportional to the binary interaction parameter χ in equation (3). The parameter χ should be inversely proportional to absolute temperature and independent of solution composition.

In practice, it is difficult to assign the number of repeating segments in solvent or polymer unambiguously. For this reason, it is usual in using Flory-Huggins theory to replace segment fraction (Ψ) in equation (3) by volume fraction (Φ). This is done by assuming that the number of segments per molecule is proportional to the molar volume of the molecule (v). Volume fraction in a solvent/polymer solution is therefore represented by

$$\Phi_1 = 1 - \Phi_2 = \frac{v_1 x_1}{v_1 x_1 + v_2 x_2} \tag{5}$$

Equation (3) therefore becomes

$$a_1 = \Phi_1 \exp\left[(1-v_1/v_2)\Phi_2 + \chi\Phi_2^2\right] \tag{6}$$

There is an unfortunate consequence of replacing segment fraction by volume fraction. For given composition (weight or mole fraction), volume fraction is temperature dependent because molar volume is temperature dependent.

Equivalent expressions for equations (3) and (6) exist for the polymer (7). The Flory-Huggins expressions can also be extended to multicomponent systems (7).

From the outset, Flory (6) and Huggins (4,5) recognized that their expressions for polymer solution thermodynamics had certain shortcomings (3). Among these were the fact that the Flory-Huggins expressions do not predict the existence of the LCST (see Figure 2) and that in practice the χ parameter must be composition dependent in order to fit phase equilibrium data for many polymer solutions (3,8).

The compositions dependence of the χ parameter is illustrated in Figure 4 for several common polymer/solvent combinations. It should be noted in considering Figure 4 that changes in χ correspond to exponential changes in solvent activity, as can be seen from equation (6). Numerous authors, but most notably Koningsveld and co-workers (9), have attempted to model empirically the composition dependence of χ. While this approach may prove useful in modelling certain results, it is neither predictive nor reliable for the wide variety of temperatures and pressures of interest to design engineers.

Free-Volume Models of Polymer Solutions.

In 1953, Prigogine et al. (10) published a paper which has led to a fundamental revision of models for polymer solution thermodynamics. The concept of Prigogine and co-workers may perhaps be summarized in the following way. The Gibbs energy of mixing, ΔG^M, can be represented as composed of two contributions: an enthalpy (or heat) of mixing contribution (ΔH^M) and an entropy of mixing contribution (ΔS^M), or

$$\Delta G^M = \Delta H^M - T\Delta S^M \tag{7}$$

Both Flory and Huggins derived statistical mechanical expressions for ΔS^M. Their expressions are still among the best available. For this reason, Prigogine and his co-workers concentrated their efforts on revising the statistical mechanical configurational partition function which leads, among other things, to ΔH^M.

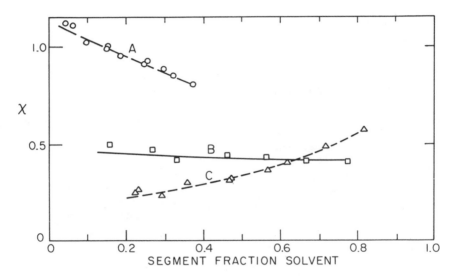

*Figure 4.   Variation of Flory–Huggins interaction parameter with composition:
Curve A—polyisobutylene/benzene at 25°C; Curve B—polyethylene/heptane at
109.9°C; Curve C—polystyrene/chloroform at 25°C.*

Prigogine's work led to a complete representation of polymer solution thermodynamics. Because of the form of Prigogine's expressions, they are often referred to as free-volume expressions.

Flory ([11]) improved the notation and form of Prigogine's expressions, and it is essentially the Flory form of Prigogine's free-volume theory that is of most use for design purposes. The Flory work ([11]) leads to an equation of state which obeys the corresponding-states principle:

$$\frac{\tilde{p}\tilde{v}}{\tilde{T}} = \frac{\tilde{v}^{1/3}}{\tilde{v}^{1/3} - 1} - \frac{1}{\tilde{v}\tilde{T}} \tag{8}$$

where $\tilde{p}$ = reduced pressure = p/p^*

$\tilde{T}$ = reduced temperature = T/T^*

$\tilde{v}$ = reduced volume = v_{sp}/v^*_{sp}

v_{sp} = volume per gram of fluid (specific volume)

The terms p^*, T^*, and v^*_{sp} are characteristic reducing parameters which may be obtained by fitting pressure-volume-temperature data (density, thermal expansion coefficient, and thermal pressure coefficient) for each pure component in the mixture ([3],[12]). Values of p^*, v^*_{sp}, and T^* are given in Tables I and II.

Equation (8) is a corresponding-states expression and therefore can be used for mixtures as well as for pure components. For mixtures, the characteristic parameters are given by ([3])

$$p^* = \Psi_1^2 p_1^* + \Psi_2^2 p_2^* + 2\Psi_1\Psi_2 p_{12}^* \tag{9}$$

and

$$T^* = p^*/(\Psi_1 p_1^* /T_1^* + \Psi_2 p_2^*/T_2^*) \tag{10}$$

The terms Ψ_i are segment fractions, defined by

$$\Psi_i = w_i v^*_{isp}/(w_1 v^*_{1sp} + w_2 v^*_{2sp}) \tag{11}$$

The term p_{12}^* is a binary interaction parameter which must be determined from phase equilibrium data. We will discuss determination of p_{12}^* values in more detail later.

The reduced volume ($\tilde{v}$) for a mixture can be obtained by solving equation (8) numerically. An approximate fit of reduced volume from equation (8) for pure components or for mixtures is given by

$$\tilde{v} = 0.99734 + 1.9644 \ \tilde{T} + 30.735 \ \tilde{T}^2 + 1.9756 \times 10^3 \ \tilde{T}^3$$

Table I.

Characteristic Parameters for Solvents

Solvent	v_{sp}^*, mL/g	p^*, atm	T^*, K
acetone	1.020	2882	5070
acrylonitrile	1.000	4422	5060
benzene	0.8900	5696	4780
carbon tetrachloride	0.4870	5587	4700
chloroform	0.5460	3168	5280
cyclohexane	1.020	5064	5060
1,2-dichloroethane	0.6560	4926	5440
diethyl ether	1.060	2478	4310
dimethyl formamide	0.9170	4255	6240
p-dioxane	0.8060	3327	5730
ethyl acetate	0.8970	4827	4820
ethyl benzene	0.9260	5301	5210
ethylene glycol	0.7760	8144	6360
n-heptane	1.140	3800	4840
n-hexane	1.198	4836	4768
methyl ethyl ketone	0.9580	5656	4590
nitrobenzene	0.6960	7019	6110
n-pentane	1.220	2626	4330
styrene	0.9230	5568	5920
2,2,4-trimethyl-pentane	1.130	4501	4840
vinyl acetate	0.8630	3820	5120
p-xylene	0.9710	4985	5930

Table II.

Characteristic Parameters for Polymers

Polymer	v^*_{sp}, mL/g	p^*, atm	T^*, K
polyacrylic acid	0.6940	8951	7270
polyacrylonitrile	0.8160	7175	28600
polybutadiene	0.8842	9870	3889
poly-n-butyl methacrylate	0.8320	6415	8010
polychloroprene	0.7020	7500	7600
polydimethylsiloxane	0.8610	3306	5100
polyethylacrylate	0.7280	5625	6580
polyethylene, high density	1.040	4342	7370
polyethylene, low density	0.9990	6415	6800
polyethylene oxide	0.7530	6632	6450
polyisobutylene	0.9540	4352	7870
cis-1,4-polyisoprene	0.9320	5152	7460
polymethyl methacrylate	0.7620	11250	11400
polypropylene, atactic	1.000	5517	6940
polystyrene, atactic	0.8170	5270	7970
polyvinyl acetate	0.7290	6869	7400
polyvinyl chloride	0.6240	8645	7960

$$-4.8443 \times 10^4 \; \tilde{T}^4 + 3.1093 \times 10^5 \; \tilde{T}^5 + 1.2355 \times 10^5 \; \tilde{T}^6$$

$$+3.7833 \times 10^4 \; \tilde{T}^7 \tag{12}$$

The Flory formulation yields an expression for solvent activity in a solvent(1)/polymer(2) solution ($\underline{3}$):

$$a_1 = \Psi_1 \exp\left\{ (1-r_1/r_2)\Psi_2 + \frac{p_1^* M_1 v_{1sp}^*}{RT_1^*} \left[3 \ln\left(\frac{\tilde{v}_1^{1/3} - 1}{\tilde{v}^{1/3} - 1} \right) \right. \right. \tag{13}$$

$$\left. \left. +1/\tilde{T}_1 (1/\tilde{v}_1 - 1/\tilde{v})] + \frac{\Psi_2^2 M_1 v_{1sp}^*}{RT\tilde{v}}(p_1^* + p_2^* - 2p_{12}^*) \right\} \right.$$

where $r_1/r_2 = M_1 v_{1sp}^* / M_2 v_{2sp}^*$

A similar expression can be derived for the polymer activity ($\underline{11}$). The formalism can also be extended to multicomponent solutions ($\underline{13}$).

Values of p_{12}^* can be determined, in principle, from any phase equilibrium data. A small table of p_{12}^* values is available in reference ($\underline{3}$). However, one of the most straight-forward ways of determining p_{12}^* values is to fit phase equilibrium data for solvent sorption in concentrated polymer solutions. To do this, equations (2) and (13) are combined to solve for p_{12}^* utilizing experimental partial pressure data.

To recapitulate, the Flory version of the Prigogine free-volume or corresponding-states polymer solution theory requires three pure-component parameters (p^*, v_{sp}^*, T^*) for each component of the solution and one binary parameter (p_{ij}^*) for each pair of components.

Application of Corresponding-States Theory.

To illustrate the application of corresponding-states theory to polymer solution calculations, we consider two cases of solvent/polymer vapor-liquid equilibria. The first case we consider is that of the chloroform/polystyrene solution. The second is that of benzene/polyethylene oxide.

The chloroform/polystyrene solution exhibits highly non-ideal behavior. As shown by curve C in Figure 4, the χ parameter for this solution rises from a low value to a high value as solvent concentration increases. However, as shown in Figure 5, the partial pressure of chloroform above a mixture of

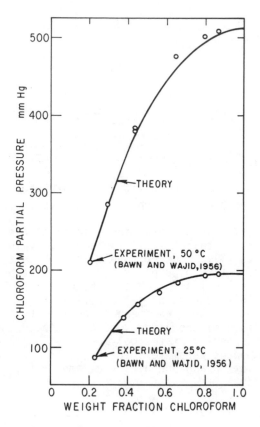

Figure 5.   Partial pressure of chloroform in polystyrene

chloroform and polystyrene (14) can be fit by the corresponding-
states model with p^*_{12} values of 4220 atm at 25°C and 4197 atm
at 50°C (3). The interaction parameters are independent of
solution composition. Such a fit with a composition independent
interaction parameter is a significant improvement over Flory-
Huggins theory.

One of the few sets of data for solvent/polymer vapor-
liquid equilibria covering a wide range of temperatures is that
of Chang (15) for benzene sorption by polyethylene oxide. Figure
6 is a plot of the natural logarithm of the interaction parameter versus reciprocal absolute temperature for the benzene/
polyethylene oxide system from 75°C to 150°C. The semilogarithmic plot in Figure 6 is linear within statistical precision
and provides an extremely good representation of the temperature
dependence of p^*_{12}. We expect such behavior to be general,
provided that there are no morphological changes in the solution
over the temperature range involved.

Gas-Polymer Equilibria.

It is often of industrial interest to be able to predict the
equilibrium sorption of a gas in a molten polymer (e.g., for
devolatilization of polyolefins). Unfortunately, the Prigogine-
Flory corresponding-states theory is limited to applications
involving relatively dense fluids (3,8). An empirical rule
of thumb for the range of applicability is that the solvent
should be at a temperature less than 0.85 T_R, where T_R is the
absolute temperature reduced by the pure solvent critical
temperature.

For application to gas sorption in polymers, we have modified the Prigogine-Flory formalism to apply to low- and high-
density fluids and their mixtures (12). The modified equation
of state has the form

$$\frac{\tilde{p}\tilde{v}}{\tilde{T}} = \frac{1}{rc} + \frac{1}{\tilde{v}^{1/3}-1} - \frac{1}{\tilde{v}\tilde{T}} \tag{14}$$

where $rc = p^* v^*_{sp} M/RT^*$. The term M is the molecular weight, and
c is one third of the number of external degrees of freedom
per segment in the molecule. Note that for completely spherical
molecules (argon, methane) r = 1 and c = 1 (corresponding to
three translational degrees of freedom). For polymer molecules
rc >> 1. There are therefore two limiting cases of equation
(14):

$$\frac{\tilde{p}\tilde{v}}{\tilde{T}} = \frac{\tilde{v}^{1/3}}{\tilde{v}^{1/3}-1} - \frac{1}{\tilde{v}\tilde{T}} \tag{15}$$

for spherical molecules, and

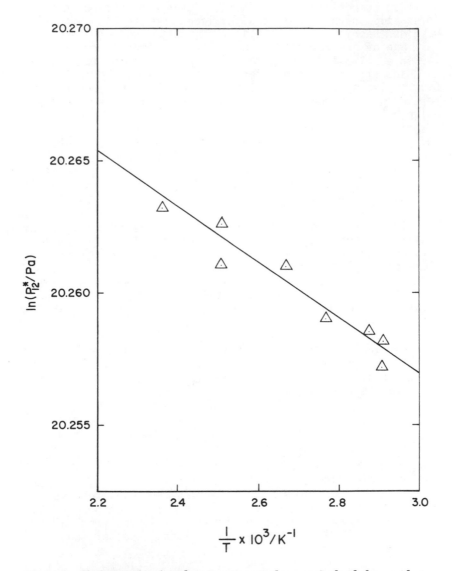

Figure 6.   Variation of $p_{12}{}^{*}$ with temperature for benzene/polyethylene oxide

$$\frac{\tilde{p}\tilde{v}}{\tilde{T}} = \frac{1}{\tilde{v}^{1/3} - 1} - \frac{1}{\tilde{v}\tilde{T}} \tag{16}$$

for pure polymers. Equation (15) for spherical molecules is identical to equation (8) of the Flory model.

The expression for solvent activity is

$$
\begin{aligned}
a_1 = \Psi_1 \exp\Bigg\{ &(1 - r_1/r_2)\Psi_2 + \frac{M_1 v^*_{1sp}}{R}\Bigg[3\frac{p^*_1}{T^*_1} \ln \frac{\tilde{v}_1^{1/3} - 1}{\tilde{v}^{1/3} - 1} \\
&+ p/T(\tilde{v}^{4/3} - \tilde{v}_1^{4/3}) + 1/T(p^*/\tilde{v}^{2/3} - p^*_1/\tilde{v}_1^{2/3}) \\
&+ p^*_1/T^*_1 - p^*/T^*\Bigg] + \left(\frac{p^*_1 M_1 v^*_{1sp}}{RT^*_1} - 1\right) \ln(\tilde{v}/\tilde{v}_1) - 1 - \\
&r_1/r(\tilde{v}^{1/3} - 1) + \tilde{v}_1^{1/3} + \frac{M_1 v^*_{1sp}}{RT}\Bigg[X_{12}\Psi_2^2/\tilde{v} \\
&- p^*_1(1/\tilde{v} - 1/\tilde{v}_1) + p^*_1/\tilde{v}_1 - p^*/\tilde{v}\Bigg]\Bigg\}
\end{aligned} \tag{17}
$$

where $X_{12} = p^*_1 + p^*_2 - 2p^*_{12}$

$$r_1/r = \Psi_1 + \Psi_2(r_1/r_2) = \Psi_1 + \Psi_2(M_1 v^*_{1sp}/M_2 v^*_{2sp})$$

There is a similar expression for polymer activity. However, if the fluid being sorbed by the polymer is a supercritical gas, it is most useful to use chemical potential for phase equilibrium calculations rather than activity. For example, at equilibrium between the fluid phase (gas) and polymer phase, the chemical potential of the gas in the fluid phase is equal to that in the liquid phase. An expression for the equality of chemical potentials is given by Cheng (12).

To illustrate the use of the gas sorption model, we show in Figure 7 results of the supercritical ethylene sorption in low-density polyethylene (12,16). As seen in Figure 7, the theory is capable of fitting the ethylene sorption data. In this instance, the data at three temperatures can be fit within experimental precision using interaction parameters (p^*_{12}) of 3235 atm, 3178 atm, or 3101 atm at 126°C, 140°C, and 155°C, respectively.

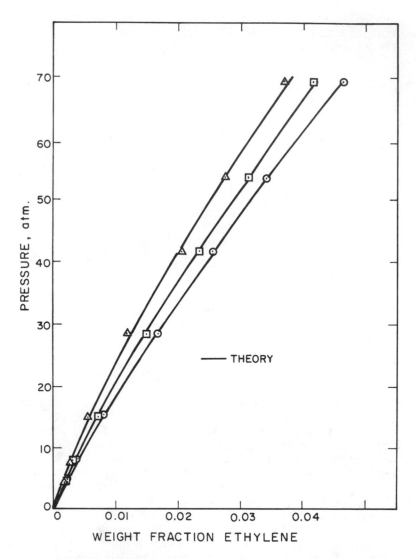

Figure 7.   Ethylene sorption in polyethylene: (◯) *126°C;* (☐) *140°C;* (△) *155°C.*

Multicomponent Mixtures - Multiple Solvents.

There are two types of multicomponent mixtures which occur in polymer phase equilibrium calculations: solutions with multiple solvents or polymers and solutions containing polydisperse polymers. We will address these situations in turn.

There are relatively few phase equilibrium data relating to concentrated polymer solutions containing several solvents. Nevertheless, in polymer devolatilization, such cases are often of prime interest. One of the complicating features of such cases is that, in many instances, one of the solvents preferentially solvates the polymer molecules, partially excluding the other solvents from interaction directly with the polymer molecules. This phenomenon is known as "gathering".

Unfortunately, relatively little work has been done on the solution thermodynamics of concentrated polymer solutions with "gathering". The definitive work on the subject is the article of Yamamoto and White (17). The corresponding-states theory of Flory (11) does not account for gathering. We therefore restrict our consideration here to multicomponent solutions where the solvents and polymer are nonpolar. For such solutions, gathering is unlikely to occur.

We have recently extended the Flory model to deal with nonpolar, two-solvent, one polymer soltuions (13). We considered sorption of benzene and cyclohexane by polybutadiene. As mentioned earlier, a binary interaction parameter is required for each pair of components in the solution. In this instance, we required interaction parameters to represent the interactions benzene/cyclohexane, benzene/polybutadiene, and cyclohexane/ polybutadiene.

It is clear that much more work should be done on this important subject.

Multicomponent Mixtures - Polymer Polydispersity.

Essentially all industrial polymers are polydisperse. The effect of polymer polydispersity on phase equilibrium has been discussed previously by many authors, but the treatment of Tompa (2) is one of the most complete. For our purposes, the situation can be summarized as follows. Polydispersity has virtually no effect on vapor-liquid equilibria (as long as the polymer is non-volatile). However, polymer polydispersity does have an important influence on liquid-liquid equilibria.

There is a large body of experimental literature relating to polymer fractionation in liquid-liquid equilibria. In addition, numerous authors have analyzed polymer fractionation using Flory-Huggins theory. We have considered use of the corresponding states theory to model polymer fractionation for the ethylene/ polyethylene system at reactor conditions (18). Results of the

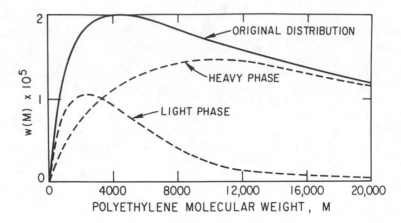

Figure 8. Fractionation of polyethylene owing to phase splitting in ethylene solution: molecular weight distributions in equilibrium phases at 260°C and 900 atm

calculations are shown in Figure 8. The algorithm used to generate Figure 8 is generally applicable to liquid/polydisperse
polymer calculations.

The algorithm we used for solvent/polydisperse polymer
equilibria calls for only one solvent/polymer interaction parameter. The interaction parameter (p^*_{12}) used in the algorithm can
be determined from essentially any type of ethylene/polyethylene
phase equilibrium data. Cloud-point data have been used (18),
while Cheng (16) and Harmony (19) have done so from gas sorption
data.

Conclusion.

We have reviewed here, in the brief space available, some
recent developments in phase equilibrium representations for
polymer solutions. With these recent developments, reliable
tools have become available for the polymer process designer
to use in considering effects of phase equilibrium properly.

The corresponding-states theory of polymer solution
thermodynamics, developed principally by Prigogine and Flory,
has provided a reliable predictive tool requiring only minimal
information. We have seen here several examples of the use of
the corresponding-states theory. We have also seen that the
corresponding-states theory is a considerable improvement over
the older Flory-Huggins theory.

Many further developments can be expected in the use of
corresponding-states polymer solution theory in engineering
practice. However, the reliability and versatility of this
method is now well demonstrated for engineering use.

Literature Cited

1. Prigogine, I., and Defay, R., "Chemical Thermodynamics",
   Longmans, Green, London, 1967.

2. Tompa, H., "Polymer Solutions", Butterworths, London, 1957.

3. Bonner, D. C., and Prausnitz, J. M., AIChE J.(1973), 19, 943.

4. Huggins, M. L., J. Chem. Phys.(1941), 9, 440.

5. Huggins, M. L., Ann. N. Y. Acad. Sci.(1942), 43, 9.

6. Flory, P. J., J. Chem. Phys.(1941), 9, 660.

7. Flory, P. J., "Principles of Polymer Chemistry", Cornell University Press, Ithaca, N.Y., 1953.

8. Bonner, D. C., J. Macromol. Sci. - Revs. Macromol. Chem.(1975), C13, 263.

9. Koningsveld, R., and Staverman, A. J., J. Polym. Sci., Part A-2(1968), 6, 305.

10. Prigogine, I., Trappeniers, N., and Mathot, V., Disc. Faraday Soc.(1953), 15, 93.

11. Flory, P. J., J. Amer. Chem. Soc.(1965), 87, 1833.

12. Cheng, Y. L., and Bonner, D. C., J. Polym. Sci. - Phys. Ed. (1978), 16, 319.

13. Dincer, S., and Bonner, D. C., Ind. Eng. Chem., Fundam.(1979), in press.

14. Bawn, C. E. H., and Wajid, M. A., Trans. Farad. Soc.(1956), 52, 1658.

15. Chang, Y. H., and Bonner, D. C., J. Appl. Polym. Sci.(1975), 19, 2457.

16. Cheng, Y. L., and Bonner, D. C., J. Polym. Sci. - Phys. Ed. (1977), 15, 593(1977).

17. Yamamoto, M., White, J. L., and McLean, D. L., Polymer(1971), 12, 290.

18. Bonner, D. C., Maloney, D. P., and Prausnitz, J. M., Ind. Eng. Chem., Proc. Des. Dev.(1974), 13, 91.

19. Harmony, S. C., Bonner, D. C., and Heichelheim, H. R., AIChE J.(1977), 23, 758.

RECEIVED February 1, 1979.

Propylene Polymerization Kinetics in Gas Phase Reactors Using Titanium Trichloride Catalyst

N. F. BROCKMEIER

Amoco Chemicals Corp., Naperville, IL 60540

Gas phase olefin polymerizations are becoming important as manufacturing processes for high density polyethylene (HDPE) and polypropylene (PP). An understanding of the kinetics of these gas-powder polymerization reactions using a highly active $TiCl_3$ catalyst is vital to the careful operation of these processes. Well-proven models for both the hexane slurry process and the bulk process have been published. This article describes an extension of these models to gas phase polymerization in semibatch and continuous backmix reactors.

This article documents the mathematical development of these gas phase kinetic models and compares the calculated results (reaction rates, yields, operating conditions) with published results (1). The correlation of these results is quite promising, enough to indicate that these models may be fully capable of describing gas phase PP kinetics. Most of the kinetic data presently available come from laboratory semibatch reactors. Probably the greatest utility of this modeling work is to provide a rigorous method to design large-scale continuous reactors from the semibatch runs performed in a laboratory. The results should be valuable to the process designer.

Mathematical Development of Models

The kinetic models for the gas phase polymerization of propylene in semibatch and continuous backmix reactors are based on the respective proven models for hexane slurry polymerization (2). They are also very similar to the models for bulk polymerization. The primary difference between them lies in the substitution of the appropriate gas phase correlations and parameters for those pertaining to the liquid phase.

The kinetic models are the same until the final stage of the solution of the reactor balance equations, so the description of the mathematics is combined until that point of departure. The models provide for the continuous or intermittent addition of monomer to the reactor as a liquid at the reactor temperature.

The monomer vaporizes instantaneously and mixes completely with
the gas in the reactor. Any mass transfer resistance to mixing
is neglected with respect to other resistances. Agitation of the
reacting powder is assumed to be sufficient to intermix it uniformly without entraining it in the gas above. The gas is assumed
to circulate through the powder sufficiently to prevent concentration gradients except within the gas trapped in a growing polymer
particle. A porous shell of polymer grows with geometric similar-
ity around the catalyst particle, which is assumed to be spherical. Scanning electron micrographs of finished powder particles
indicate that the original catalyst particle disintegrates under
certain conditions. The assumption is that polymer grows concen-
trically around each of the fragments. Propylene must diffuse
through tortuous passages in this shell containing a stagnant mixture of inert gases such as saturated hydrocarbons, since diffusion of propylene through solid polymer is much too slow to contribute to the reaction. The propylene reaches the active catalyst surface, where it reacts at concentration C_s, which is generally somewhat lower than C_E, the concentration in the well-mixed
gas phase. The heat transfer resistance between polymerizing
solids and gas has been neglected, so both solids and gas are at
the same temperature (1). The semibatch reactor operates with
monomer feed on pressure control with no materials leaving. The
continuous reactor has feeds of catalyst and monomer, and powder
removal to hold a constant level.

The model postulates two significant resistances in series:
diffusion through the growing shell (R_{DF}) and polymerization at
the catalyst surface (R_{CAT}). The catalytic reaction resistance,
R_{CAT}, is intended to include any and all of the effects of the
sorption rate of monomer on the surface, steric arrangement of
active species, the addition of the monomer to the live polymer
chain, and any desorption needed to permit the chain to continue
growing. We assume a steady state in which every mole of propylene that polymerizes is replaced by another mole entering the
shell from the gas, so that all of the fluxes are equal to N_V
gmol propylene reacted per second per liter of total reactor
volume. The following set of equations relates the molar flux
to each of the concentration driving forces.

$$N_V = k_c A_c \ (C_E - C_s) \quad \text{Diffusion through} \qquad (1)$$
$$\text{porous shell}$$

$$N_V = k_s A_c \cdot C_s \qquad \text{Catalyst surface} \qquad (2)$$
$$\text{reaction}$$

The catalyst surface area is defined in the following relationship (3):

$$A_c = \frac{6X_m}{\rho_7 d_7 V_R} \qquad (3)$$

Note that the flux and the area A are based on unit reactor volume. This permits direct comparison between resistances during the course of a reaction because it remains constant. Propylene concentration is expressed in gmol per liter of gas, a number which is kinetically significant. The activity of the propylene contacting the catalyst surface is assumed to be proportional to its concentration at the surface, C_s.

The series nature of the model permits calculation of the overall reaction resistance (R_{OV}) simply by summing the individual resistances:

$$R_{OV} = R_{CAT} + R_{DF} \tag{4}$$

Equations 1 and 2 are rearranged to eliminate C_s and put into the form of equation 4, in which $R_{OV} = C_E/N_V$:

$$\frac{C_E}{N_V} = \frac{1}{k_s A_c} + \frac{1}{k_c A_c} \tag{5}$$

The solution of equation 5 for flux N_V provides the kinetics we desire. Numerous experiments with the hexane slurry system have led to the development of an expression for k_c that is partly based on theory and partly on an empirical constant in the denominator:

$$k_c = \frac{D_{AB}}{0.0245 \cdot d_7 \cdot Y_t} \tag{6}$$

Experience has shown that the mass transfer rate decreases as the reciprocal of Y_t as polymerization proceeds ($\underline{2}$). We assume that this is the same for both slurry and gas phase polymerization. The diffusivity, D_{AB}, is estimated by a method recommended for gases at high pressure. The method used is derived from equations of Mathur and Thodos ($\underline{4}$):

$$D_{AB} = \frac{5.43 \times 10^{-5} \cdot T \cdot v \cdot P_{cm}^{2/3}}{T_{cm}^{7/6} \cdot \overline{M}_w^{1/2}} \tag{7}$$

The value of v is important both in equation 7 and for accurate calculation of concentrations in other equations. For simplicity and accuracy, the Peng-Robinson equation of state has been used to calculate v for the model ($\underline{5}$). This equation expresses the P-V-T relationship as follows:

$$P = \frac{RT}{v - b} - \frac{a(T)}{v(v+b) + b(v-b)} \tag{8}$$

For equations 7 and 8, temperatures must be in degrees Rankine. The quantities a(T) and b are defined in the literature (5).

Refer again to equation 5 -- the value of C_E can now be calculated from v:

$$C_E = x_2/62.43v \qquad (9)$$

where x_2 = mole fraction propylene in vapor and the 62.43 factor converts the units to g-mol/ml. The remaining quantity in equation 5 is the rate constant k_s. Much experience with slurry polymerization has resulted in the following equation to describe how the rate constant decays with the age of a catalyst particle in the reactor and how it increases with an increase in temperature:

$$k_s = 0.992 \times 10^9 \exp (-E_A/RT) \cdot k_s° \exp (-t/\lambda) \qquad (10)$$

where the superscript on k_s denotes the original value at 80°C. The second exponential in equation 10, exp $(-t/\lambda)$, is the same as that reported by Wisseroth for his studies of gas phase propylene polymerization (1). His parameter b equals $1/\lambda$. The value of λ is sensitive to temperature, so we assume that:

$$\lambda = \lambda' B \exp (-E_\lambda/RT) \qquad (11)$$

where the prime denotes the value at 80°C.

A more generally useful rate constant is defined by the following equation:

$$r = k_k \cdot X_m \cdot C_s \cdot M_A \qquad (12)$$

The often-quoted instantaneous catalyst activity is $r \div X_m$ in g/g-hr. The value of k_k is always proportional to k_s, according to the following:

$$k_k = \frac{6 \cdot 3600 \cdot k_s}{1000 \cdot \rho_7 \cdot d_7} \qquad (13)$$

This now is the point of departure at which the semibatch treatment follows a different course from treatment of a continuous reactor.

Semibatch Model "GASPP". The kinetics for a semibatch reactor are the simpler to model, in spite of the experimental challenges of operating a semibatch gas phase polymerization. Monomer is added continuously as needed to maintain a constant operating pressure, but nothing is removed from the reactor. All catalyst particles have the same age. Equations 3-11 are solved algebraically to supply the variables in equation 5, at the desired operating conditions. The polymerization flux, N_v, is summed over three-minute intervals from the startup to the desired residence time, τ, in hours:

$$W_p = \sum_{t=0.05}^{t=\tau} (3600 \cdot 0.05 \cdot N_V V_R M_A) \qquad (14)$$

to give W_p, the cumulative production of polymer. All parameters that are functions of time, such as k_s, k_k, Y_t, and N_V are placed in this loop that sums the polymer production in three-minute intervals. This interval is sufficiently small that it behaves as an infinitesimal, so the summation in equation 14 is equivalent to integration. The computer output normally prints the values of all important parameters at one hour intervals. The user may change this for his convenience.

Continuous Model "CØNGAS". This model predicts performance of an ideal continuous wellstirred polyreactor. The model system consists of a continuous backmix reactor in which the total powder volume is held constant. There are four inlet streams: 1) Makeup of pure propylene, 2) Catalyst feed, 3) Hydrogen feed, and 4) Recycle. The single effluent powder stream is directed through a perfect separator that removes all solids and polymer and then the gases are recycled to the reactor. The makeup propylene is assumed to disperse perfectly in the well-mixed powder.

An arbitrary decision was made to fix the mass of catalyst in the reactor, rather than the feed rate of catalyst. The feed rate is calculated from the loading and the mean residence time:

$$X_{mf} = X_m / \tau \qquad (15)$$

The polymerization rate in the reactor in g/hr is calculated from:

$$r = 3600 \cdot N_V \cdot M_A \cdot V_R \qquad (16)$$

The yield of solid polymer per g of $TiCl_3$ is:

$$Y_t = r/X_{mf} \qquad (17)$$

The yield of polymer is assumed to be the sum of the insoluble and soluble polypropylene. The basis for this simple formulation of the yield and rate is grounded in the following relationship:

$$Y_{t,c} = M_A \cdot C_s(t) \cdot k_k(t) \cdot \tau \qquad (18)$$

For all likely operating conditions, (ie., for $\tau < \lambda$), the appropriate values of the concentration and the polymerization rate constant are the values calculated at $t = \tau$ (2). To prove this, the exit age distribution function for a backmix reactor was used to weight the functions for C_s and k_k and the product was integrated over all exit ages (6). It is enlightening at this point to compare equation 18 with one that describes the yield attainable in a typical laboratory semibatch reactor at comparable conditions.

$$Y_{t,s} = M_A \cdot \int_0^t C_s(t)\, k_k(t)\, dt \qquad (19)$$

The yield that can be attained by a semibatch process is generally higher because the semibatch run starts from scratch, with maximum values of both variables: $C_s(o) \cong C_E$ and $k_k(o) = k_k{}^\circ$. However, the yield from a continuous run in which τ equals the batch time is governed by the product of $C_s(\tau)$ and $k_k(\tau)$, so $Y_{t,c} \leq Y_{t,s}$. The penalty in yield attainable in a continuous run can be eliminated by two routes. If $R_{DF} = 0$, $C_s(t)$ will be constant; and if the catalyst does not deactivate, $\lambda \to \infty$ and $k_k(t) = k_k{}^\circ$. Because neither of these conditions is likely to be fulfilled completely, a continuous polymerization in a backmix reactor will probably always fail to attain the Y_t attainable by a semibatch reactor at the same τ. However, several backmix reactors in series will approach the behavior of a plug flow continuous reactor, which is equivalent to a semibatch reactor.

Refer to equation 5, which relates N_V to the parameters in the reactor. For the continuous reactor these parameters are evaluated at $t = \tau$. However, the solution to equation 5 is complicated by the fact that N_V is not only on the left hand side, but N_V also appears in the expression for R_{DF} as a first power. Newton's method of convergence is used to solve equation 5 for the continuous reactor.

Experimental

The experimental semibatch apparatus and procedure have been described in several places through the text of Wisseroth's publications (1, 7-9), so the details will not be repeated here. For nearly all of his work the reactor volume was one liter, temperature was 80°C, pressure was 30 atm (441 psia), and the feed was polymerization grade C_3H_6. I assume that the reactor gas composition was 99% C_3H_6 and 1% inerts. The range of catalyst loading was from 11 to 600 mg of $TiCl_3$ per batch. The reaction time was varied from 0.5 to 6 hours. The weight ratio of alkyl-to-$TiCl_3$ in the catalyst recipe was varied from 0.5 to 32. No data are reported from a continuous gas phase reactor.

According to Wisseroth, the agitator design was quite important, and was very similar to those shown in reference 1. The
speed was adjustable from 0-360 rpm and a gland packing seal was
used. For special operations, metallic balls were added to the
reactor to improve temperature stability (10).

Discussion of Results

This section is divided into three parts. The first is a
comparison between the experimental data reported by Wisseroth
(1) for semibatch polymerization and the calculations of the kinetic model GASPP. The comparisons are largely graphical, with
data shown as point symbols and model calculations as solid curves. The second part is a comparison between some semibatch reactor results and the calculations of the continuous model CØNGAS.
Finally, the third part discusses the effects of certain important process variables on catalyst yields and production rates,
based on the models.

Semibatch Simulation, GASPP. The experimental results in
Tables 3 and 4 of reference 1 appear to fall into three groups
of different activity for the BASF $TiCl_3$ used. Figure 1 shows a
group of runs with the lowest catalyst activity, most of which
had a catalyst recipe with an alkyl/$TiCl_3$ (Alk/Ti) ratio of 8:1
or 16:1. Figure 1 shows the course of semibatch polymerization
of propylene at 80°C and 441 psia for reaction times of from 1
to 4 hours. The family of curves shows that the total (insoluble
plus soluble) polymer formed is directly proportional to the mass
of $TiCl_3$ charged and that it increases with time at a gradually
decaying rate. The slope of a curve at any point is the instantaneous rate. The data points are quite scattered, probably because of gas phase experimental difficulties. The family of model
curves was adjusted to the best visual fit to the data (esp.load-
ings of 100 and 30 mg $TiCl_3$) by varying the initial rate constant
$k_s°$ keeping the characteristic lifetime, λ, constant at 11.1 hr.
Diffusion resistance, R_{DF}, is not very important for this low
activity catalyst. Thus, the curvature of these model curves is
nearly all catalyst rate decay. Wisseroth claims that this catalyst has a decay parameter, b, equal to 0.09 hr.$^{-1}$ (essentially
the same as $1/\lambda$) (1).
Wisseroth's Tables 3 and 4 also include data for catalysts
of much higher activity. These more active catalysts tend to be
those for which the Alk/Ti ratio is lower, ie. 2:1 or 1:1. However, there are exceptions to this tendency in all three groups.
Figure 2 shows his rate data for the two groups of higher activity catalysts, along with solid model curves for all three groups.
All results are shown as yield in grams of PP per gram of $TiCl_3$
loading for ease in comparison. The model curves were adjusted
to the best visual fit using only $k_s°$, keeping all other parameters constant. The more active catalysts have values of $k_s°$
larger than the base activity by factors of 2.2 and 8.6, respec-

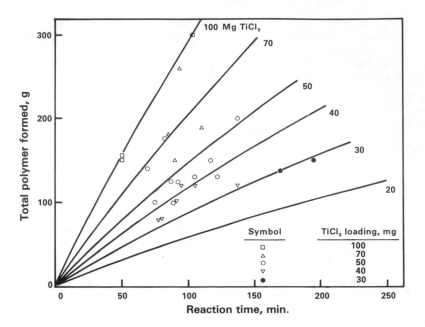

Figure 1.   Gas Phase Propylene Polymerization (batch reactor, 1 L, 80°C, 441 psia, 99% pure C₃H₆)

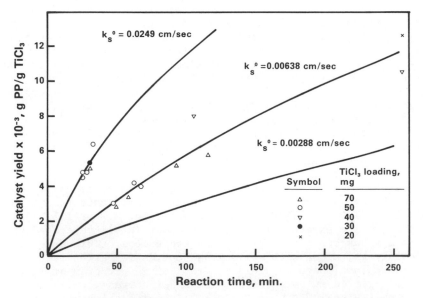

Figure 2.   Gas Phase Propylene Polymerization with catalysts of various activities (batch reactor, 1 L, 80°C, 441 psia, 99% pure C₃H₆)

tively (10). The effects of R_{DF} are definitely noticeable with
the two higher activity catalysts. These two upper curves in
Figure 2 demonstrate a significant concave downward curvation,
showing that R_{DF} retards the polymerization rate in the same
manner as catalyst decay. Lack of data for longer reaction times
at the highest activity prevents a more quantitative conclusion
about R_{DF} estimation in the model. The fit for the middle curve
at 4+ hours looks promising. This good fit is especially note-
worthy because the empirical constant of 0.0245 in equation 6 is
the same value as that used in the slurry model CØNTPP (2).

The output from the semibatch model GASPP permits a detailed
look at the way in which polymerization resistances increase during the course of a run. Figure 3 shows how these resistances
increase with yield for a run using the intermediate activity
catalyst (k_s° = 0.00638 cm/sec), with 50 mg TiCl$_3$ loading. Overall resistance, R_{OV}, is the sum of the individual resistances in
series, $R_{CAT} + R_{DF}$. The curve for R_{CAT} increases with an upward
curvature because the catalyst activity decays with time. Yield
increases with time as in Figure 2. The equations indicate that
for constant composition, R_{DF} is proportional to yield, Y_t, as
shown by Figure 3. For 80°C, 441 psia, and 99% C$_3$H$_6$, the equation for BASF catalyst is

$$R_{DF} = 3.38 \times 10^{-6} \cdot Y_t/X_m \qquad (20)$$

independent of catalyst k_s°. A change in k_s° only shifts R_{CAT}
up or down, and of course shifts R_{OV}. For the catalyst modelled
in Figure 3, R_{DF} is about half of R_{CAT} at Y_t = 12,000, so the
polymerization rate is largely under kinetic control at high
yield. However, for the most active catalyst, R_{CAT} drops to
about 1/4 of the present value and becomes the lesser contribution to R_{OV}. Then polymerization becomes diffusion-controlled
at high yields. For a 30% drop in operating pressure, the constant in equation 20 drops 40%. A 9°C drop in temperature, however, hardly affects the constant in equation 20, all other things
constant. For comparison of resistances between runs with different catalyst loadings, note that the product of resistance
and X_m is a constant for a given catalyst (for instance, in Figure 1).

There are very sparse data available at the long residence
times that are needed to evaluate the characteristic lifetime (λ)
of the BASF TiCl$_3$ used by Wisseroth (1). Figure 4 shows these
few values of $\bar{A}$ (mean activity) for the intermediate activity
catalyst at 80°C, covering a range of from 1 to 6 hours. For convenience on this semilog plot, all information has been normal-
ized by dividing it by the original value at zero run time. The
constant used for A° is 3710 g/g hr. The straight line and two
curves in Figure 4 were generated with model GASPP for comparison
with the data. The solid curve fits the experimental data for $\bar{A}$
satisfactorily. If there were little or no mass transfer limi-

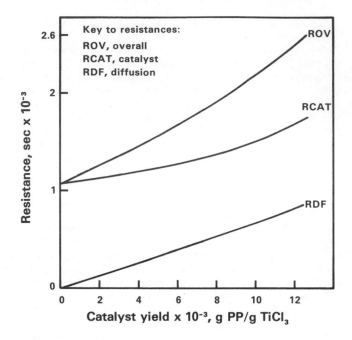

Figure 3.   Resistance increase during semibatch gas phase propylene polymerization (80°C, 441 psia, 50 mg TiCl₃ loading, 99% pure C₃H₆, $k_s^\circ = 0.00638$ cm/sec)

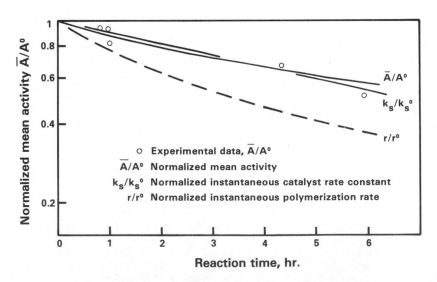

Figure 4.   Decay of catalytic activity (propylene gas phase polymerization 80°C, 441 psia)

tations to the kinetics, the mean activity would decay more slowly than the rate constant k_S. This is observed with calculations
on the catalyst of lowest activity. However, this intermediate
catalyst has sufficient diffusion limitation to cause the decay
of $\bar{A}$ to nearly match the decay of k_S, as shown in Figure 4. This
happens because diffusion causes r (dashed curve) to drop so
rapidly with time. This coincidence of the curves for $\bar{A}$ and k_S
means that the experimenter might easily miss seeing any diffusion mechanism in operation, were it not for other evidence (higher activity catalyst, other temperatues, or slurry systems). From
the slope of the straight line for k_S decay, λ is 9.68 hr. at
80°C. Wisseroth reports b equals to 0.09 hr.$^{-1}$, or λ = 11.1 hr.
at 80°C. This might be the conclusion from the slope of a
straight line drawn through the curved data $\bar{A}$, even though this
is not a rigorously correct way to evaluate b.

The semibatch model GASPP is consistent with most of the
data published by Wisseroth on gas phase propylene polymerization.
The data are too scattered to make quantitative statements about
the model discrepancies. There are essentially three catalysts
used in his tests. These BASF catalysts are characterized by
the parameters listed in Table I. The high solubles for BASF are
expected at 80°C and without modifiers in the recipe. The fact
that the BASF catalyst parameters are so similar to those evaluated earlier in slurry systems lends credence to the kinetic
model.

Continuous Simulation, CØNGAS. There are no published data
available on propylene continuous polymerization suitable to
check the accuracy of the CØNGAS model. However, there is an
equation for yield vs. time published by Wisseroth ($\underline{1}$) for a
completely backmixed continuous reactor:

$$Y_t = \frac{A° \cdot \bar{\tau}}{1 + b\,\bar{\tau}} \tag{21}$$

where $\bar{\tau}$ = mean residence time in reactor, hr. I assume the reactor is perfectly backmixed for this discussion. The CØNGAS
model develops yield vs. time as an output, but there is no
simple expression such as equation 21. For comparison between
CØNGAS and equation 21, the R_{DF} must be set equal to zero. When
this is done, the yields calculated by CØNGAS average about 4%
lower than the yields from equation 21 over the range $0 \le 4 \le \bar{\tau}$.
This 4% discrepancy is much less than the typical experimental
variations of 20% or more, so it seems reasonable to assume that
the CØNGAS model is accurate enough for design use. It has been
developed in the same way as the other well-proven Amoco PP kinetic models.

Figure 5 is a plot of the calculated polymer yields from the
continuous model CØNGAS vs. the yields from the semibatch model

Table I

BASF CATALYST PARAMETERS AT REFERENCE TEMPERATURE OF 80°C
GAS PHASE PROPYLENE POLYMERIZATION

Recipe	Rate Const. k_s°, cm/s	Decay Time λ, hr.	Characteristic Size, d_7 Microns	Rate Constant, liter/hr-g $TiCl_3$ k_k°
8/1	0.0249	9.68	9.	265.
2/1	0.00638	9.68	9.	67.7
2/1	0.00288	9.68	9.	30.5

NOTE:

Recipe refers to wt. ratios of $Al(C_2H_5)_3/TiCl_3$.

For all catalyst: E_A = 14,500 cal/mole ([11]) and E_λ =
      -3,735 cal/mole.

Wt. Percent Solubles: 20-30.

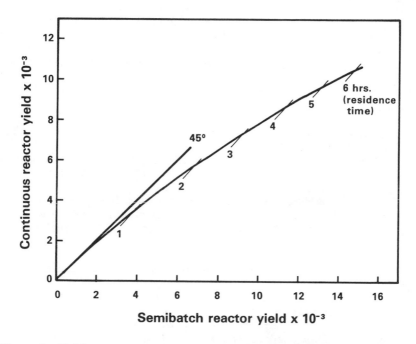

Figure 5. Yield comparison between semibatch and continuous polyreactors
(80°C and 441 psia, $k_s° = 0.00638$ cm/sec, 99% pure C_3H_6)

GASPP, at the same conditions of $\underline{80}°C$, 441 psia, $k_s° = 0.00638$ cm/sec, and at the same value of $\bar{\tau}$. Wisseroth's equations ($\underline{1}$) give the same result. The 45° line indicates the locus of equal yields. Obviously, the comparison between reactor yields for this BASF $TiC\ell_3$ catalyst in the gas phase system is essentially the same as for the many other catalysts tested in slurry and bulk ($\underline{2}$). At the same $\bar{\tau}$, the yield for a continuous backmix reactor is always less than for a semibatch reactor. The relative values of the instantaneous rates are just the reverse of the yields (Fig. 6 in ref. $\underline{2}$). This yield penalty arises because the catalyst activity decays and because R_{DF} is significant. These factors operate on the RTD in a continuous reactor to reduce the yield. Figure 5 shows a 25% yield reduction in a continuous reactor at $\bar{\tau} = 6$ hr, based on the model calculations. Tests with various $k_s°$ values give qualitatively similar curves using the models. If the yield penalty is much less than 25%, this could indicate that the RTD is more characteristic of plug flow, that R_{DF} is very small, or that the decay rate is very small. The yield penalty can be reduced by staging backmix reactors in series.

The CØNGAS simulation was used to generate the three yield vs. time profiles in Figure 6 using the most active BASF catalyst at 60°, 80°, and 100°C. At 60°C, diffusion becomes dominant only at the higher yields, whereas at 80°C and 100°C, all of the results are diffusion-dominated ($R_{DF} \leq R_{CAT}$). The diffusion effects reduce the slopes of the two upper profiles to about one-half. From 80°C, a 20° boost in reaction temperature causes the yield to increase by 16%, while a 20°C drop causes yield to decrease by about 37%.

Reactor Variable Study. Assuming that the kinetic models are valid, we have a means to rapidly explore the effects of making certain changes in the catalyst or in the operating conditions. Fortunately, Wisseroth published the results for two runs at 100°C and two more runs at 20 atm in his Table 3 ($\underline{1}$). The model GASPP was used to correlate yield vs. time for the 20°C boost to 100°C reaction temperature. With the first run, a value of $k_s° = 0.00198$ cm/sec was required to achieve the low yield reported. His second run had a yield of 13750 at 4.68 hr. Model GASPP requires $k_s° = 0.00294$ cm/sec to give this result at 100°C. This rate constant is only 2% greater than the $k_s°$ reported in Table I here for the lowest activity BASF $TiC\ell_3$. On this basis, I will assume that these kinetic models correctly account for temperature changes. More data are needed to verify this. The temperature effect in GASPP is practically the same as that claimed by Wisseroth in a recent letter ($\underline{10}$).

Model GASPP was also used to correlate the results for polymerization at 20 atm, a 33% reduction in reactor pressure. Using the parameters for the most active BASF $TiC\ell_3$, the model yields were 13% and 40% higher than the experimental yields. The 13% is

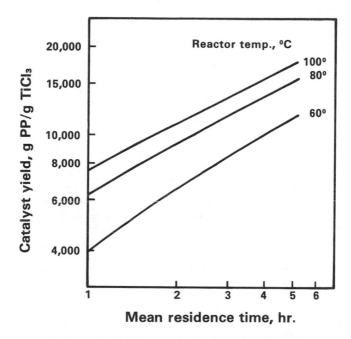

Figure 6. Simulation of a continuous backmix reactor (propylene gas phase polymerization—$k_s° = 0.0249$ cm/sec, $\lambda = 9.68$ hr. 400 psia; reactor gas composition—99% C_3H_6, 1% inerts)

certainly within experimental variation. The activity of the
other run at least falls between the activities tabulated for
BASF TiCl$_3$. The two 20 atm runs were terminated at 0.53 and 0.70
hours, respectively. There are well-known difficulties in
accurately determining the kinetics from such short runs. Although these results are insufficient to draw a conclusion, the
model response to pressure is consistent with the data.

Table II summarizes the yields obtained from the CØNGAS computer output variable study of the gas phase polymerization of
propylene. The reactor is assumed to be a perfect backmix type.
The base case for this comparison corresponds to the most active
BASF TiCl$_3$ operated at almost the same conditions used by
Wisseroth, 80°C and 400 psig. Agitation speed is assumed to
have no effect on yield provided there is sufficient mixing. The
variable study is divided into two parts for discussion: catalyst
parameters and reactor conditions. The catalyst is characterized
by k_s°, λ, and d_7. Percent solubles is not considered because
there is presently so little kinetic data to describe this. The
reactor conditions chosen for study are those that have some
significant effect on the kinetics: temperature, pressure, and
gas composition.

The base case is listed in the second column of Table II,
for $1 < \tau < 5$ hrs. The increase in yield with time is quite similar
to curves shown in Figure 6, in which there is a 50% increase in
yield as τ is boosted from one to two hours. The yield increases
only 25% more with an additional hour of reaction time. Consider
the effect of doubling the value of k_s°. (For all except one case,
all other parameters are kept constant.) The yield at τ = 3 hr.
is boosted by only 11%, clearly demonstrating that this model of
continuous polymerization is strongly diffusion-controlled at
these conditions with this catalyst. If a catalyst is developed
that has approximately double the lifetime, τ = 18.3 hr., the
yields will grow as shown in the fourth column. This change im-
proves the base yield at 3 hr. by about 2%. Greater catalyst
stability at reactor conditions is of little benefit to this
process. The fifth column shows how to change parameters so as
to keep k_k° constant, reducing only the diffusion resistance,
R_{DF}. The characteristic size of the TiCl$_3$, d_7, is reduced by 45%.
The 3-hr. yield is thereby increased by 53%, a very significant
benefit to the process. Methods to achieve this kind of change
are well worth investigation.

The last three columns in Table II demonstrate changes in
reactor conditions, using the same catalyst. A 9°C drop in
temperature causes an 11% drop in yield at τ = 3 hr. A 33% drop
in pressure causes only a 14% drop in yield. These small yield
changes are expected because the system is diffusion-controlled.
Composition changes have some intricate effects on the kinetics.
The propylene diffusivity in the gas mixture will depend on composition, as the inerts content is changed. These changes are
felt in R_{DF}, and the changes in yield might be significant for a
diffusion-controlled reaction. Another effect is simple dilution

Table II

REACTOR VARIABLE STUDY

Yields from a Continuous Backmix Reactor, Simulated with CØNGAS

Mean Residence Time, τ, hr.	Base Case[1] Yields	Catalyst Parameters			Operating Conditions		
		Rate Constant $k_s°$ 0.0524 cm/sec	Lifetime, λ=18.3 hr.	Size d_7= 0.0005 cm (+k_s=0.01384)	Temp., T=71°C	Pressure P=261.8 psig	Composition C_3H_6,Inerts .76, .24
1	6180	7180	6250	8940	5270	5020	5870
2	9290	10470	9450	14020	8190	7860	8400
3	11630	12960	11900	17820	10390	9980	10400
4	13530	15020	13940	20870	12190	11680	12060
5	15160	16810	15710	23400	13710	13100	13490

NOTE 1:

Base Case $k_s° = 0.0249$ cm/sec. λ=9.68 hr.

$d_7 = 0.0009$, $ρ_7 = 2.26$,

$T_{op} = 80°C$, $P_{ga} = 400$ psig, $X_m = 0.05$ g $TiCl_3$

Composition, mole fraction in vapor. C_3H_6: 0.99,   Inerts: 0.01

of the monomer by the inerts that might accumulate in a continuous process with recycle. The rate equation (No. 12) is first-
order in monomer concentration. The last column in Table II
shows a case with about 25% lower propylene concentration. The
yield is reduced by 11%, once again showing the modifying effect
of diffusion-control. The conclusion is that polypropylene polyreactors tend to be diffusion-controlled, whether the process is
slurry, bulk, or gas phase. The difference is in the yields
achieved before diffusion begins to control the reaction.

ABSTRACT

Appropriate equations for propylene polymerization with $TiCl_3$
catalyst in a gas phase system are assembled into complete mathematical simulations for both a semibatch and a continuous backmix polyreactor. These simulations are an extension of the well-
proven hexane slurry kinetic models with the substitution of gas
phase equations for the liquid phase. The semibatch model
(GASPP) is verified as an accurate model by testing with data
published by BASF, whereas the continuous model (CØNGAS) is derived from GASPP using the equations appropriate to backmixing.
Given the inputs of catalyst parameters such as activity, stability, and particle size and operating conditions such as temperature, pressure, reaction time, and gas composition, these models
generate yield and production rate as outputs. The models are
estimated to have less than 5% error for the following range of
conditions:

Temperature:	60° to 100°C
Pressure:	50 to 450 psia
Reaction Time:	0 to 6 hours
Catalyst Activity:	22 to 597 liter/hr-g $TiCl_3$

These models indicate that propylene gas phase polymerization
with a highly active $TiCl_3$ catalyst shifts from kinetic control
at short reaction times to diffusion control at longer times as
the catalyst yield exceeds about 4000 g.PP/g.$TiCl_3$. Measures to
reduce this limitation would significantly benefit the process.
The effects of diffusion and catalyst decay cause yields from a
continuous backmix reactor to be 25 to 30% lower than from a semibatch reactor at the same residence time. This yield penalty can
be reduced by staging backmix reactors in series.

NOTATION

A   Catalyst activity, g./g./hr.

A_c Catalyst surface area per unit volume of reactor, cm^2/liter.

B   Proportionality constant, 0.0025.

C   Concentration, moles/liter.

D_{AB} Diffusivity of propylene in propane, cm^2/sec.

d   Diameter, cm.

E_A Activation energy for polymerization, 14,500 cal/mole (11).

E_λ Activation energy, -3735 cal/mole.

k   Mass transfer coefficient or rate constant, cm./sec.

M_w Molecular weight g./g-mole.

N   Propylene flux in g-mole/sec-liter.

P   Pressure, lb./sq.in.abs.

R   Mass transfer or other resistance, sec.

r   Polymerization rate, g./hr.

T   Temperature, °K.

t   Mean age of reactor contents, hr.

V   Volume, liter.

v   Mixture specific volume, ft^3/lb-mol.

W_p Cumulative polymer, g.

X_m Catalyst loading, g.TiCℓ_3.

x   Mole fraction in vapor.

Y_t Yield, g.polypropylene/g.TiCℓ_3.

λ Catalyst characteristic time, hr.

ρ Density, g./cm^3.

τ Exit age of reactor fluid, hr.

SUBSCRIPTS

A   Propylene.

B   Propane and inerts.

c   Variable is defined in terms of concentration difference.

cm   Refers to pseudocritical mixture.

CAT   Catalyst surface.

DF   Diffusion in polymer shell passages.

E   Equation of state value.

k   Volumetric rate constant, liter/hr.-g $TiCl_3$.

m   Mixture.

mf   Mass feed rate.

OV   Overall.

s   Surface of catalyst.

v   Per unit volume.

7    $TiCl_3$ catalyst.

ACKNOWLEDGEMENT

    The author thanks Amoco Chemicals Corp. for permission to publish this manuscript. The publications of Dr. K. Wisseroth have provided a vital input to verify this mathematical development.

LITERATURE CITED

1. Wisseroth, K., Chemiker Zeitung, (1977), 101, 271.
2. Brockmeier, N. F. and Rogan, J. B., "Simulation of Continuous Polymerization Processes", AIChE Symp. Ser. No. 160, (1976), 72, 28.
3. Satterfield, C.N. and Sherwood, T. K., "The Role of Diffusion in Catalysis", pp. 45-47, Addison-Wesley, Reading, Mass., 1963.
4. Mathur, G. P. and Thodos, G., AIChE J., (1965), 11, 613.
5. Peng, D. Y., and Robinson, D. B., I & E. C. Fundam. , (1976), 15, 59.
6. Levenspiel, O., "Chemical Reaction Engineering", pp. 112-116 and Ch. 9, Wiley, New York, 1962.
7. Wisseroth, K., Angew. Makromol.Chemie, (1969), 8, 41.
8. Wisseroth, K., Kolloid Z. and Z. Polym., (1970), 241, 943.
9. Wisseroth, K., Chemiker-Zeit.,(1973), 97, 181.
10. Wisseroth, K., personal communication, Jan. 6, 1978.
11. Natta, G. and I. Pasquon, "Advances in Catalysis", Vol. II, pp. 21-23, Academic Press, N.Y., 1959.

RECEIVED January 15, 1979.

Free-Radical Polymerization: Sensitivity of Conversion and Molecular Weights to Reactor Conditions

KIU H. LEE and JOHN P. MARANO, JR.[1]

Union Carbide Corporation, P. O. Box 8361, South Charleston, WV 25303

One on-going objective in a commercial polymerization reactor of a fixed size is to maximize the reactor productivity at the desired product properties. Polymerization reactors are sensitive to changes in operating parameters because the reactors involve highly exothermic reactions. A relatively minor fluctuation in the operating variables could cause wide fluctuations in the reactor responses. Therefore, it is important to search out underlying relationships concerning the reactor performance and the modes of reactor operations. The conceptual trends obtained in this type of investigation provide valuable information regarding the operating limits of a given reactor and also aid in determining the future actions aimed at further improving the limits of the reactor performance. The conceptual study may dictate changes in the initiator system, the solvent system (chain transfer agents) and the heat transfer system for a reactor of fixed size to provide the maximum possible conversion at desired product properties.

The study of the peak temperature sensitivity to the reactor operating parameters and the construction of sensitivity boundary curves for stable reactor operation were previously reported (1). This paper presents a computer study on conceptual relationships between the conversion-product properties and the reactor operating parameters in a plug flow tubular reactor of free radical polymerization. In particular, a contour map of conversion-molecular weight relationships in a reactor of fixed size is presented and the sensitivity of its relationship to the choice of initiator system, solvent system and heat transfer system are discussed.

In the study, the kinetic rate constants applicable to the polymerization of ethylene (2,3) were used with an assumed activation volume. These values appear to be a reasonably consistent set of constants for the polymerization of ethylene and, as shown

Current address: [1]Now located at Mobil Chemical Company, Edison, New Jersey.

0-8412-0506-x/79/47-104-221$07.75/0

in Figure 1, generate a temperature profile which appears to be reasonably typical of the one measured in a high-pressure ethylene polymerization reactor.

This work particularly emphasizes the importance of selecting the initiator system for optimum reactor operation and reveals general concepts which specify the desired properties and operational modes of an optimum initiator system. In addition, the effects of the system heat transfer and the CTA (chain transfer agent) level on the conversion-molecular weights relationships are presented.

Polymerization Tubular Reactor Model

The computer model used for this analysis is based on a plug flow tubular reactor operating under restraints of the commonly accepted kinetic mechanism for polymerization reactions ($\underline{3},\underline{4}$). The computer model consists of the numerical integration of a set of differential equations which conceptualizes the high-pressure polyethylene reactor. A Runge-Kutta technique is used for integration with the use of an automatically adjusted integration step size. The equations used for the computer model are shown in Appendix A.

The elements of the model are the reaction mechanism, heat and mass balance equations, and the molecular weight moment equations, which are numerically integrated with reactor length. The molecular weight moment equations were derived using moment generating functions ($\underline{5}$). This method of derivation appears to be the most reliable technique for deriving molecular weight moments. The assumptions used in the computer program are listed in the Appendix. But unlike many polymerization models, no assumptions are made concerning the steady-state concentration of radicals since the radicals will not be at steady-state under conditions of rapidly changing temperature over the entire range of reactor conditions which must be considered in this analysis.

The computer model used in this analysis was discussed previously ($\underline{1},\underline{6}$) and are similar, in general concepts, to other models ($\underline{7},\underline{8}$) discussed in the literature. The computer program was written for use on IBM 370/65 computer.

Use of Reactor Model. In order to begin the study of the sensitivity of the reactor responses, the simplest reactor configuration possible was chosen. This paper considers the case of constant pressure, constant heat transfer coefficient and constant jacket or wall temperature, with initiation occurring by a free radical generator which decomposes by a first-order rate process. The efficiency of the initiators considered is assumed constant (=0.5) and, as with the initiator efficiency, all other rate constants are assumed independent of viscosity. Following this initial investigation, an optimization study of reactor configurations is planned, lifting most of these initial restrictions.

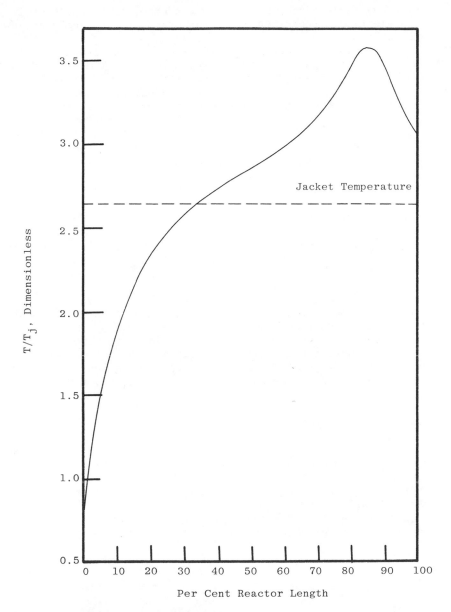

Figure 1. Typical reactor temperature profile for continuous addition polymerization: a plug-flow tubular reactor. Kinetic parameters for the initiator: $I_o = 10$ ppm; $E_d = 32.921$ kcal/mol; ln $k_d' = 26.492$ ln sec^{-1}; $f = 0.5$. Reactor parameter: $[(4hT_j\tau)/(D\rho C_p)] = 5148.2$. [$(C_p) =$ heat capacity of the reaction mixture; $(\rho) =$ density of the reaction mixture; $(h) =$ overall heat-transfer coefficient; $(T_j) =$ reactor jacket temperature; $(\tau) =$ reactor residence time; $(D) =$ reactor diameter].

The fixed variables used in the computer simulation are shown in Table 1 along with the kinetic rate constants for the polymerization reactions.

Since it is a conceptual study employing a theoretical reactor model, it is also important to appreciate the limits of this type of investigation. The advantage of the computer investigation over a pilot or production reactor investigation is the obvious cost and time saving over the real reactor experiment. The computer investigation can also yield a more definable relationship with fewer parameter excursions since the output will be free of scatter. In addition, excursions in reactor parameters can be taken which might be considered unsafe on or beyond the equipment limitations of an existing real reactor.

The pitfalls of a computer model are obvious in that it is only a conceptual representation of the reactor and includes only as many aspects of the real reactor as present knowledge permits. In addition, even the most perfectly conceived description will still depend upon the accuracy of the physically measured constants used in the model for the quality of the process representation. The goal of this report is, however, only to show conceptual trends and the technological base is developed to the extent that the conceptual trends will be correct. In some respects the computer model is a better process development tool than the pilot plant used for the LDPE process since the pilot reactor does not yield directly scaleable information. The reader should take care to direct his attention to the trend information and conceptual differences developed in this work; very little attention should be paid to the absolute values of the parameters given.

Theoretical Considerations

The overall reactions involved in a free radical polymerization are described in the Appendix. It is interesting however, to look into several reaction steps which contain the key reaction parameters and control the rate of production and the molecular weights of the polymer.

For illustration of some simple highlights, the rate of polymerization is given by the relationship

$$R_p = k_p[M][R^*] \tag{1}$$

the rate of radical generation is given by the relationship

$$R_R = R_i - 2k_t R^{*2} \tag{2}$$

and the rate of initiation is given by the relationship

$$R_i = 2fk_d[I] \tag{3}$$

TABLE I

REACTION AND REACTOR PARAMETERS USED IN THE COMPUTER SIMULATION

KINETIC RATE CONSTANTS

	Log Base e Frequency Factor (1, mole, sec)	Activation Energy/R (°C)	Activation Volume/R (°C/atmo)
Pll	17.891	3573.0	-0.2800
TC	20.796	150.0	-0.1710
T	1.198	150.0	-0.1100
CFM	-4.616	1988.4	0.0366
CFP	-4.963	304.3	0.0366
CFS	-0.577	1476.2	0.0366
BETA*	27.0	13940.0	0.244

PHYSICAL PARAMETERS

BMO = 17.176   PP = 33,000   CP = 16.2399   BM = 28.0
HEATRO = 22,300.00   TF = 58.0   DMO = 17.176

Pll = propagation rate
 TC = termination by combination rate constant
  T = ratio of the termination rate constant for combination
      to the rate constant for disproportionation
CFM = ratio of the rate constant for monomer transfer to the
      constant for propagation
CFP = ratio of the rate constant for polymer transfer (long-
      chain branching) to the constant for propagation
CFS = ratio of the rate constant for solvent transfer to the
      constant for propagation
BMO = inlet monomer (ethylene) concentration, mole/l
 PP = reactor pressure, psia
 BM = monomer (ethylene) molecular weight
 CP = heat capacity at constant pressure of the reaction
      fluid, cal/mole-°C
HEATRO = heat of reaction for the polymerization, cal/mole
   TF = reactor inlet temperature, °C
  DMO = reactor fluid density, mole/l
 BETA = β-scission reaction rate constant

*Though this reaction is important in LDPE reactors, it was ignored
in the present simulation because of the uncertainty of the rate
constant value and for simplification aimed at representing trends.

Further, the major molecular weight controls are accomplished
in the rate of chain transfer by solvent (R_{ts}) and the rate of
radical terminations (R_{tr}) by combination and disproportionation.

$$R_{ts} = k_{ts}[s][R*] \qquad (4)$$

$$R_{tr} = k_{tr}[R*]^2 \qquad (5)$$

where

k_p = propagation rate "constant"
k_t = termination rate "constant"
k_d = first order rate "constant" for the initiator
     breakdown
$[M]$ = monomer concentration
$[I]$ = initiator concentration
$[R*]$ = radical concentration
f = initiator efficiency
k_{ts} = chain transfer rate "constant"
k_{tr} = termination (combination or disproportionation)
     rate constant
$[s]$ = solvent (chain transfer agent) concentration

Equation (1) shows the rate of polymerization is controlled
by the radical concentration and as described by Equation (2) the
rate of generation of free radicals is controlled by the initiation rate. In addition, Equation (3) shows this rate of generation is controlled by the initiator and initiator concentration.
Further, the rate of initiation controls the rate of propagation
which controls the rate of generation of heat. This combined
with the heat transfer controls the reaction temperature and the
value of the various reaction rate constants of the kinetic mechanism. Through these events it becomes obvious that the initiator
is a prime control variable in the tubular polymerization reaction
system.

Further, the rate constants may be written in its useful form
as

$$k = k''e^{-E/RT} e^{-\Delta VP/RT} \qquad (6)$$

where

E = activation energy
R = ideal gas constant
T = absolute temperature
P = pressure
ΔV = activation volume
k'' = frequency factor

The values of k'' and E are highly dependent on the initiator types
and their effects on the solvent types are less overwhelming. The
types of solvent used as chain transfer agent are usually fixed

for a given reactor and only the concentration of the solvents are varied to control the molecular weights.

The initiator types, however, are characterized by these parameters, and since the effect of pressure is small ($\underline{1},\underline{9}$) and the tubular polymerization of ethylene is undertaken within a narrow range of pressure, the descriptive constant becomes

$$k_d = k'_d e^{-E_d/RT} \tag{7}$$

Any initiator which decomposes by a first-order rate process can, therefore, be characterized by the two parameters k'_d and E_d. Such important materials as organic peroxides, azo compounds, as well as many other types of materials, are described by the first-order process and as such follow the general development given in this work. The efficiency will be assumed constant and the same for all initiators with 0.5.

In a case of where the radical steady-state assumption can be made, the reactor heat balance can be written in dimensionless form at constant pressure ($\underline{1}$)

$$dT'/dZ = \beta X (1-Y)^{\frac{1}{2}} e^Q (T'-1/T') - \gamma(T'-1) \tag{8}$$

where

$$\beta = \left\{ \frac{\Delta Hk'_p M_o I_o^{\frac{1}{2}} e^Q \tau}{\rho C_p T_j} \left[\frac{fk'_d}{k'_t} \right]^{\frac{1}{2}} \right\}$$

$$\gamma = \left[\frac{4h\tau}{D\rho C_p} \right]$$

$Q = \alpha/T_j$
$\alpha = E_p + (E_d - E_t/2)/R$
$T' = T/T_j$
$Z = L/L_o$

and

T = temperature
τ = residence time
M_o = monomer inlet concentration
I_o = initiator inlet concentration
f = initiator efficiency
ρ = density
C_p = heat capacity
T_j = jacket or wall temperature (The jacket temperature when h is defined as an overall heat-transfer coefficient; the inside wall temperature when h is defined as an heat-transfer coefficient.)
h = heat-transfer coefficient
D = reactor diameter

and $\qquad$ X = monomer conversion
$\qquad\qquad$ Y = initiator conversion
$\qquad\qquad$ ΔH = heat of reaction
$\qquad\qquad$ R = ideal gas constant
$\qquad\qquad$ L = reactor length
$\qquad\qquad$ L_O = total reactor length
$\qquad$ k'_p, E_p = frequency factor and activation energy
$\qquad\qquad\qquad$ for chain propagation
$\qquad$ k'_d, E_d = frequency factor and activation energy
$\qquad\qquad\qquad$ for initiator breakdown
$\qquad$ k'_t, E_t = frequency factor and activation energy
$\qquad\qquad\qquad$ for radical termination

Equation (8) provides a general relationship between the reactor temperature profile and the operating parameters. In relating the system heat transfer to the conversion-molecular weights relationship for a reactor of fixed size, the heat transfer coefficient emerges as the correlating parameter.

Effects of Initiator Concentration and Jacket Temperature.
The ability to manipulate reactor temperature profile in the polymerization tubular reactor is very important since it directly relates to conversion and resin product properties. This is often done by using different initiators at various concentrations and at different reactor jacket temperature. The reactor temperature response in terms of the difference between the jacket temperature and the peak temperature ($\theta = T_p - T_j$) is plotted in Figure 2 as a function of the jacket temperature for various inlet initiator concentrations. The temperature response not only depends on the jacket temperature but also, for certain combinations of the variables, it is very sensitive to the jacket temperature.

The conversion reflects the temperature response realized in the reactor and the temperature response shown in Figure 2 can be replotted in terms of conversion responses and they are shown in Figure 3. The figure clearly shows that the conversion in a reactor of fixed size depends on both the inlet initiator concentration and the jacket temperature. There exist optimum operating conditions to maximize the conversion in a reactor of fixed size. The dashed lines in Figures 2 and 3 not only indicate the optimum operating condition but also show the limits of stable reactor operation for a given initiator system in a fixed reactor. In addition, the average polymer molecular weights that are produced in the reactor depend on the temperature response and thus, are related to the conversion. An example of this relationship is shown in Figure 4.

Optimum Operating Line. The relationships between the conversion and the average molecular weight can be plotted as a function of initiator concentration while varying the jacket temperature to optimize the conversion. The relationships are shown in

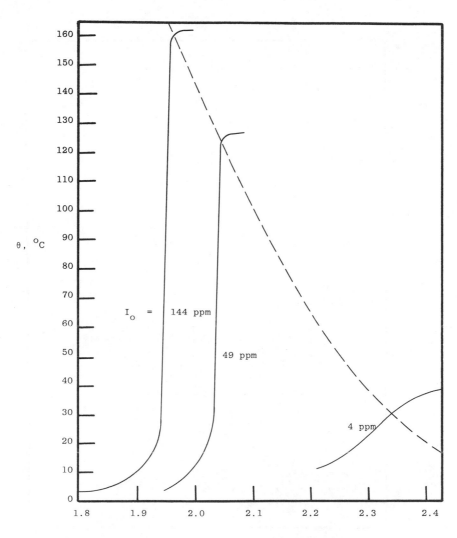

Figure 2. Effect of jacket temperature on the maximum temperature in a plug-flow reactor (f = 0.5; ln k_d' = 43.2261 ln sec^{-1}; E_d/R = 23481.06 °C; $\tau/\rho C_p$ = 0.67898 sec-ft³-°C/BTU; T_o = 58°C)

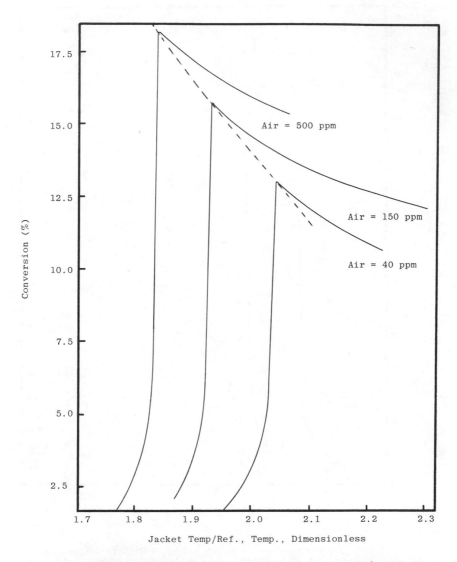

Figure 3.   Conversion–jacket temperature relation (computer simulation). Heat transfer coefficient: 75 cal/m² sec °C

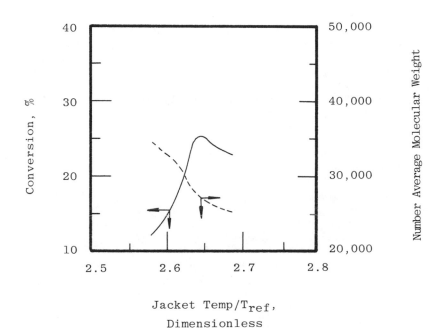

Figure 4.   Operation of a plug-flow tubular addition polymerization reactor of fixed size using a specified free-radical initiator (initiator kinetic parameters: $E_d = 32.921$ Kcal/mol; ln $k_d' = 26.492$ ln sec^{-1}; $f = 0.5$; 10 ppm initiation, 1.0 mol % solvent)

Figure 5 and it provides a basis for constructing the optimum operating line. The figure shows that there is a maximum conversion that can be achieved at a given initiator concentration. In terms of molecular weight, no optimization exists. However, an operational limitation forces an optimization since the operation of the reactor at molecular weights which are higher than can be obtained at optimum conversion could result in an excessive initiator concentration at the reactor exit and an excessive number of free radicals in the polymer recovery system.

This operational limitation coupled with the maximum conversion specification results in a defined optimal operational mode which is unique to a given initiator type. The optimum operating line is illustrated in Figure 5 as the dashed line. This optimum operating line can now be plotted for different initiator types at various solvent concentrations and heat transfer conditions to compare initiator types and to study the reactor responses to the operating parameters. The operating line shows that for a given initiator type there is a maximum molecular weight and maximum conversion which can be produced in a reactor of fixed size. The operating line serves as a sound basis for comparing the performance of the reactor as the various initiators are used and further provides the direction of search for optimum initiator system for a given product in a reactor of fixed size.

Effect of Solvent Concentration. The optimum conversion-molecular weight curve can be divided into two zones: one chain transfer controlled and the other initiator controlled. In the upper molecular weight or chain transfer controlled region changes in the initiator concentration significantly change the reactor conversion but have little effect on the molecular weight. As seen in Figure 6, however, it is in this region that the chain transfer agent has its largest effect. In the low molecular weight or initiator controlled region changes in the initiator concentration alter the molecular weight but have little effect on the reactor conversion.

A further examination of a single operating line indicates that the reactor is molecular weight limited because of an inverse relationship of the initiator concentration and the jacket temperature. The limiting molecular weight is approached as the inlet initiator concentration approaches zero and the jacket temperature approaches a limiting value dictated by the initiator type. The molecular weight at that point is given simply by the ratio of the propagation to monomer and solvent chain-transfer rate constants evaluated at the limiting reaction temperature.

On the other hand, the limiting conversion in a reactor of fixed size is dependent on the temperature and the radical concentration in the reactor and results from a predominating radical-radical interaction precipitated by an increased initiator concentration and the accompanying temperature excursion. At this point the solvent concentrations have little effect on the molecular

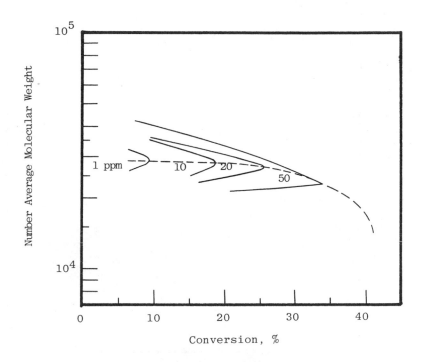

Figure 5.   Molecular weight–conversion contour map for various concentrations of a free-radical initiator operating in a tubular-addition polymerization reactor of fixed size. Curves were constructed using varying jacket temperatures (kinetic parameters for the initiator: $E_d = 32.921$ Kcal/mol; $\ln k_d' = 26.494$ ln sec^{-1}; $f = 0.5$; (– – –) optimum operating line)

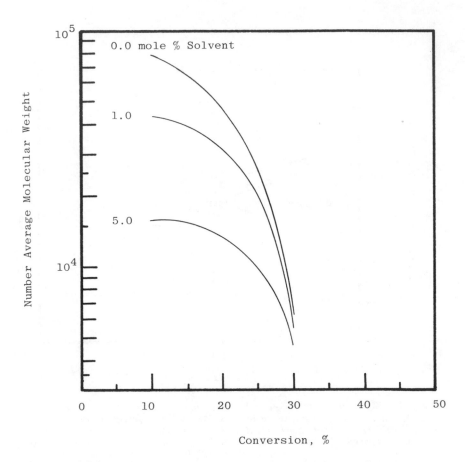

Figure 6.   Effect of solvent concentration on the molecular weight–conversion relationships of a tubular-addition polymerization reactor of fixed size using a specified initiator type. Each point along the curves represents an optimum initiator feed concentration–reactor jacket temperature combination. (kinetic parameters of the initiator: $E_d = 24.948$ Kcal/mol; ln $k_d' = 26.494$ ln sec^{-1}; $f = 0.5$)

weight. The limiting condition is reached when the chain radicals which are formed immediately terminate. This condition is dependent on the type of initiator used in the reactor.

Effects of Initiator Parameters. Initiator types can best be characterized by the frequency factor (k'_d) and the activation energy (E_d), and the effect of these parameters on the molecular weight-conversion relationship is shown in Figures 7 and 8. The curves shown are the result of choosing the jacket temperature-inlet initiator concentration combination which maximizes the reactor conversion for each initiator type investigated.

Figure 7 shows the limiting maximum molecular weight of products from a reactor of fixed size varies directly with the frequency factor of the initiator at a fixed activation energy, while the limiting conversion varies inversely with the frequency factor. In addition, the length of the chain-transfer controlled zone is increased inversely with the frequency factor.

Figure 8 shows the limiting maximum molecular weight of products produced in a reactor of fixed size varies inversely with the activation energy of the initiator at a fixed frequency factor, while the limiting conversion varies directly with the activation energy. In addition, the length of the chain-transfer controlled zone increases directly with the activation energy.

Theoretically, as the initiator activation energy approaches zero, a very high molecular weight material will be produced at a very small conversion and as the initiator activation approaches infinity, a very low molecular weight material will be produced at very high conversion. This implies that an optimum combination of E_d and k'_d which produces an infinite range of molecular weights does not exist. There is, however, an optimum combination of E_d and k'_d for a given product (given molecular weight) produced in a given reactor.

In addition to the number average molecular weight of the produced polymer, the breadth of the molecular weight distribution has important effects on the product properties. Figures 9 and 10 show the effect of initiator type on the molecular weight distribution of the resin as defined by the ratio of the weight to number average molecular weight. The figures show that the breadth of molecular weight distribution varies inversely with the activation energy of the initiator at any given conversion for an initiator of specified frequency factor and varies directly with the frequency factor for an initiator of specified activation energy. This suggests that the molecular weight distribution of a resin can be made to assume any desired value by a proper choice of the initiator.

The initiator usage can play a role in the economics of resin production. The computer simulations show the usage to be dependent on the initiator type. The effect of the initiator type on the amount of initiator required to produce a given quantity of resin at optimum reactor conditions is shown in Figures 11 and 12.

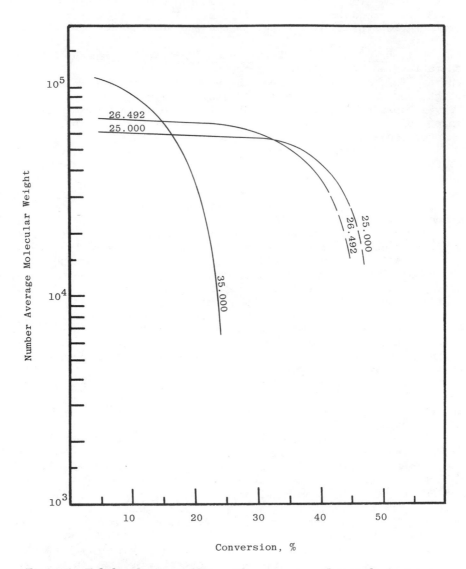

Figure 7. Tubular plug-flow addition polymer reactor: effect of the frequency factor (k_d') of the initiator on the molecular weight–conversion relationship at constant activation energy (E_d). Each point along the curves represents an optimum initiator feed concentration–reactor jacket temperature combination and their values are all different. ($E_d = 32.921$ Kcal/mol; ln $k_d' = 35.000$ ln sec^{-1}; 0.0 mol % solvent)

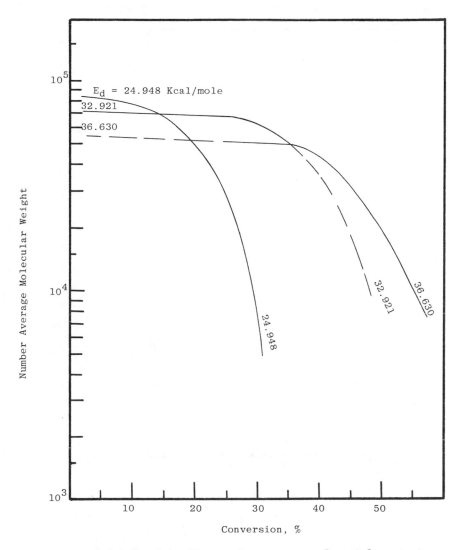

Figure 8. Tubular plug-flow addition polymer reactor: effect of the activation energy (E) of the initiator on the molecular weight–conversion relationship at constant frequency factor (k'). Each point along the curves represents an optimum initiator feed concentration–reactor jacket temperature combination and their values are all different. (ln k' = 26.494 ln sec⁻¹; 0.0 mol % solvent)

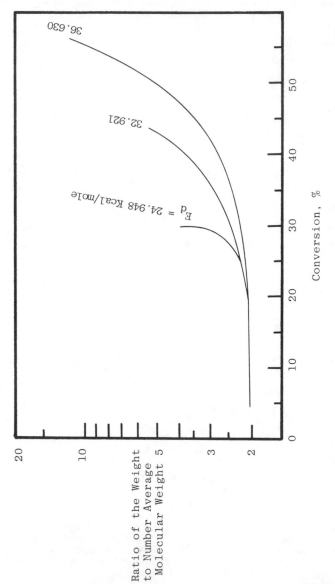

Figure 9. Effect of the initiator activation energy on the molecular weight distribution of an addition polymer produced in a tubular reactor: constant frequency factor and at widely different values of initiator–jacket temperature combination (the conversion is optimized: $\ln k_d' = 26.492$ $\ln sec^{-1}$; 0.0 mol % solvent)

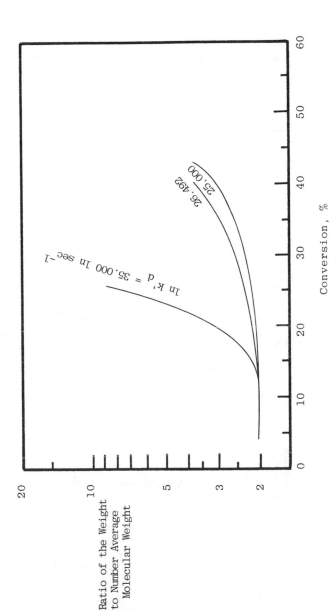

Figure 10. Effect of the initiator frequency factor on the molecular weight distribution of an addition polymer produced in a tubular reactor: constant activation energy and at widely different values of initiator–jacket temperature combination (the conversion is optimized: $E_d = 32.921$ kcal/mol; 0.0 mol % solvent)

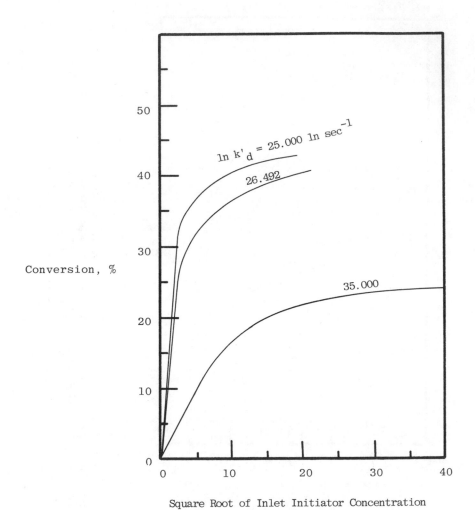

Figure 11.   *Effect of the initiator frequency factor on the initiator usage in an addition polymerization reactor: constant activation energy (the conversion is optimized; $E_d = 32.921$ kcal/mol)*

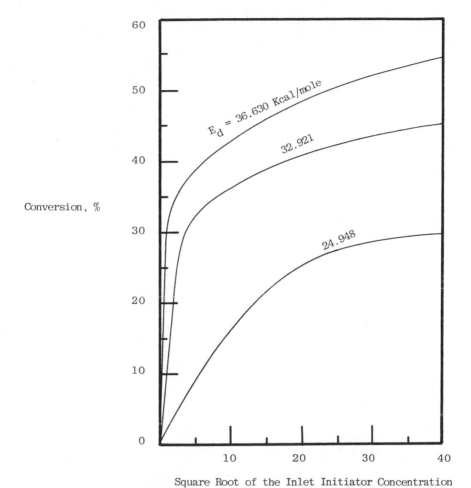

Figure 12. Effect of the initiator activation energy on the initiator usage in a tubular-addition polymerization reactor: constant frequency factor (the conversion is optimized; ln k′ = 26.492 ln sec⁻¹)

As seen in the figures, the quantity of initiator required to
yield a given conversion varies directly with the frequency factor
for an initiator with a specified activation energy and inversely
with the activation energy for an initiator of specified frequency
factor. The relationships do not show a linear proportionality
between the reactor conversion and square root of the inlet
initiator concentration.

Effect of Initiator Change on Conversion Improvement. Based
on these discussions, it is apparent that a selected initiator can
allow conversion improvements for a specified molecular weight.
This can be illustrated in Figure 13. If product A' is using
initiator A, conversion A' would result. A switch to initiator B
would cause the product D' to be produced. The initiator concentration could then be increased along curve B to product A' at
conversion A". This simple increasing of the initiator concentration and, therefore, the conversion could not have been done with
the original initiator since a decrease in the molecular weight
would have occurred. In other words a 0.1 melt index material
can be produced at the same rate as a 10 melt index material by
using an initiator of the proper design. The conversion improvement for C' product from C' conversion to C" conversion is now
done by the reversing of the initiator types and shows the sensitivity of the product properties on conversion improvement with
initiator changes.

The LDPE reactor is sometimes termed heat transfer limited
in conversion. While this is true, the molecular weight (or melt
index)—conversion relationship is not since this work shows that
a selected initiator can allow conversion improvements to be made
under adiabatic conditions for a specified molecular weight. The
actual limitation to conversion is the decomposition temperature
of the ethylene and given that temperature as a maximum limitation,
an initiator (not necessarily commercial or even known with present initiator technology) can be found which will allow any product to be made at the rate dictated by this temperature. Con-
ceptually, this is a constant (maximum) conversion reactor, running
at constant operating conditions where the product produced dictates the initiator to be used.

Effect of Heat Transfer. Because the reactor is heat transfer limited, efforts are often made to improve the heat transfer
and conversion. However, for a given initiator system in a specified reactor, there are also unique conversion-molecular weight-
heat transfer relationships. Figure 14 shows a relationship
between the average molecular weight and the conversion with heat
transfer coefficient as a parameter. The curve is based on optimized conversion-jacket temperature relationships for different
number average molecular weights. The shape of the curve implies
a given initiator system and reactor configuration. The shape of
curve may change with different reactor systems, but it does show

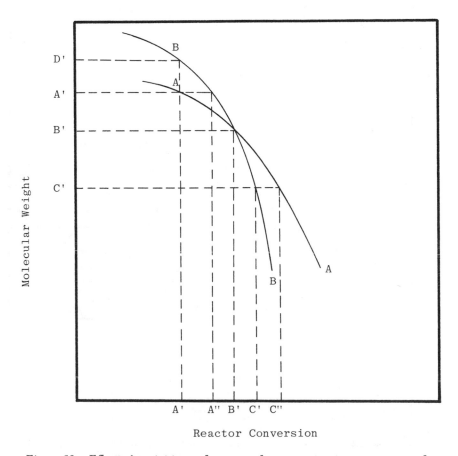

Figure 13.   Effect of an initiator change on the conversion improvement in the tubular-addition polymerization reactor

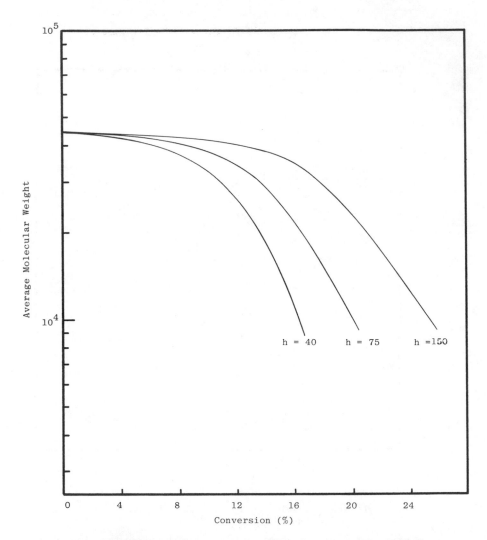

Figure 14. Molecular weight–conversion relationship (computer simulation—reactor of a fixed geometry for a given initiator system) (h) heat transfer coefficient in cal/m².°C.sec

a general trend of what can be expected of molecular weight-conversion-heat transfer coefficient relationship. The figure shows that the conversion does increase with increasing heat transfer but the degree of increase depends on the average molecular weights of polymer being produced. At high average molecular weights (lower melt index), the lines representing different heat transfer levels converge. This implies that the conversion improvement due to heat transfer is small at high molecular weight. At lower molecular weights (higher melt index), the degree of conversion improvement is much larger. In order to utilize the improvement in heat transfer for a given product, therefore, an initiator system must be selected to provide a maximum conversion spread with increasing heat transfer.

Concluding Remarks

The computer simulation study of the operation of the tubular free radical polymerization reactor has shown that the conversion and the product properties are sensitive to the operating parameters such as initiator type, jacket temperature, and heat transfer for a reactor of fixed size. The molecular weight-conversion contour map is particularly significant and it is used in this paper as a basis for a comparison of the reactor performances.

The type of initiator used affects the molecular weight and conversion limits in a reactor of fixed size and the molecular weight distribution of the material produced at a given conversion level. The initiator type also dictates the amount of initiator which is necessary to yield a given conversion to polymer, the operating temperature range of the reactor and the sensitivity of the reactor to an unstable condition. Clearly, the initiator is the most important reaction parameter in the polymer process.

The full utilization of improved heat transfer in a given reactor can only be made when the molecular weight-conversion relationships are carefully studied with various initiator types at different heat transfer levels. Then a particular initiator system must be selected for a maximum conversion improvement for a specified product.

A study of this kind can be further extended to develop optimum reactor configurations which are needed to produce given products at the highest possible conversion.

Abstract

A theoretical polymerization tubular reactor model was used to study the effects of reactor operating parameters on conversion

APPENDIX A

EQUATIONS FOR A PLUG FLOW POLYMER TUBULAR REACTOR WITH BRANCHING KINETICS

		Rate Constant
Initiation	$I_n \longrightarrow 2\,R_c$	k_{d_n}
	$R_c + M \longrightarrow R_1$	k'_d
Propagation	$R_1 + M \longrightarrow R_2$	k_p
	$R_i + M \longrightarrow R_i + 1$	
Monomer Transfer	$R_i + M \longrightarrow P_i + R_1$	k_{f_m}
Solvent Transfer	$R_i + S_n \longrightarrow P_i + R_1$	$k_{f_{sn}}$
Long Chain Branching	$R_i + P_J + R_J \longrightarrow R_J + P_i + R_i$	k_{f_p}
Termination:		
Combination	$R_i + R_J \longrightarrow P_{i+J}$	k_{t_c}
Disproportionation	$R_i + R_J \longrightarrow P_i + P_J$	k_{t_d}
Short Chain Branching	$R\text{-}(CH_2)_5\text{-} \longrightarrow R\text{-}\overset{|}{C}H\text{-}C_4H_9$	k_{SCB}
β-Scission	$R_i\text{-}C\text{-}R_k \longrightarrow R_i + P_{j+k}$	k_β
	$\overset{\displaystyle R_j}{} \searrow R_k + P_{i+j}$	

where

I_n = initiator n
S_n = solvent n
R_c = initiator fragment
R = radical of chain length i or J
M = monomer
P = polymer of chain length i or J
k = reaction rate constant of the step

Mass Balance and Molecular Weight Equations

Using the above mechanism the molecular weights are given by:

$$M_n = X_n M' = M'H'(1,t)/H(1,t) \tag{A-1}$$

$$M_w = X_w M' = M'\left\{\frac{H''(1,t)}{H'(1,t)} + 1\right\} \tag{A-2}$$

where

$$H(1,t) = \sum_1^\infty R_i + P_i \tag{A-3}$$

$$H'(1,t) = \sum_1^\infty i(R_i + P_i) \tag{A-4}$$

A-2

$$H''(1,t) = \sum_1^\infty i(i-1)(R_i + P_i) \tag{A-5}$$

and where

M_n = number average molecular weight
M_w = weight average molecular weight
M' = molecular weight of a monomeric unit

For a plug flow tubular reactor

For radicals J $(J \geq 2)$

$$\frac{dR_J}{dt} = k_pMR_{J-1} - k_pMR_J - R_J(k_{f_m}M + \sum_1^N k_{f_{sn}}S_n) - k_tR_J\sum_1^\infty R_i$$

$$- k_{f_p}(R_J\sum_1^\infty(iP_i + iR_i) - (JP_J + JR_J)\sum_1^\infty R_i) \tag{A-6}$$

For $(J \geq 4)$ add $(-2k_\beta R_J + k_\beta R_{J+3})$ to (A-6)

For polymer J $(J \geq 2)$

$$\frac{dP_J}{dt} = (k_{f_m}M + \sum_1^N k_{f_{sn}}S_n)R_J + \tfrac{1}{2}k_{t_c}\sum_1^{J-1} R_iR_{J-i} + k_{t_d}R_J\sum_1^\infty R_i$$

$$+ k_{f_p}(R_J\sum_1^\infty(iP_i + iR_i) - (JP_J + JR_J)\sum_1^\infty R_i) \tag{A-7}$$

For radical 1

$$\frac{dR_1}{dt} = \Omega - k_pMR_1 + (k_{f_m}M + \sum_1^N k_{f_{sn}}S_n)\sum_1^\infty R_i$$

$$- (k_{f_m}M + \sum_1^N k_{f_{sn}}S_n)R_1 - k_tR_1\sum_1^\infty R_i \tag{A-8}$$

$$-k_{f_p}(R_1\sum_1^\infty(iP_i + iR_i) - (P_1 + R_1)\sum_1^\infty R_i)$$

For polymer 1

$$\frac{dP_1}{dt} = (k_{f_m}M + \sum_1^N k_{f_{sn}}S_n)R_1 + k_{t_d}R_1\sum_1^\infty R_i$$

$$+ k_{f_p}(R_1\sum_1^\infty(iP_i + iR_i) - (P_1 + R_1)\sum_1^\infty R_i) \tag{A-9}$$

A-3

where $t = Z\tau$
             Z = fractional reactor length
             τ = reactor residence time
             k_p = reaction rate constant for propagation
             k_{f_p} = reaction rate constant for polymer branching
             k_{f_m} = reaction rate constant for monomer transfer
             $k_{f_{sn}}$ = reaction rate constant for solvent transfer
             k_{f_c} = reaction rate constant for termination by
                    combination
             k_{t_d} = reaction rate constant for termination by
                    disproportionation
             $k_t = k_{t_c} + k_{t_d}$
             Ω = rate of initiation

$$\Omega = \sum_1^{M''} 2f_n k_{d_n} I_n \tag{A-10}$$

             I_n = initiator concentration
             f_n = initiator efficiency
             k_{d_n} = rate constant for the breakdown of the
                    initiator
             M'' = number of initiators

$$\frac{dH(1,t)}{dt} = \Omega + k_p M C G(1,t) - \tfrac{1}{2}k_{t_c}G^2(1,t) + 2k_\beta G(1,t) \tag{A-11}$$

$$\frac{dH'(1,t)}{dt} = \Omega + k_p M G(1,t)[C + 1] - 4k_\beta G(1,t) \tag{A-12}$$

$$\frac{dH''(1,t)}{dt} = 2k_p M G'(1,t) + k_{t_c}G'(1,t)^2 + 24k_\beta G(1,t) \tag{A-13}$$
$$-12k_\beta G'(1,t)$$

where $C = C_{f_m} + (\sum_1^N C_{f_{sn}}S_n/M)$

and $C_{f_m} = k_{f_m}/k_p$
         $C_{f_s} = k_{f_s}/k_p$

$$G(1,t) = \sum_1^\infty R_i \tag{A-14}$$

$$\frac{dG(1,t)}{dt} = \Omega - k_t G^2(1,t) \tag{A-15}$$

A-4

$$\frac{dG'(1,t)}{dt} = \Omega + k_p MG(1,t) + k_p MC[G(1,t) - G'(1,t)]$$

$$-k_t G(1,t)G'(1,t) - k_{f_p}[H'(1,t)(G'(1,t) - G(1,t))\qquad (A-16)$$

$$- G(1,t)H''(1,t)] - k_\beta G'(1,t) - 2k_\beta G(1,t)$$

The integration of equations (A-11) to (A-13) and (A-15) and (A-16) along the reactor length in combination with equation (A-1) and (A-2) will give the molecular weight.

The mass balances for other components are given by

monomer:
$$- \frac{dM}{dt} = k_p MG(1,t) \qquad (A-17)$$

solvent:
$$- \frac{dS_n}{dt} = k_{f_{sn}} S_n G(1,t) \qquad (A-18)$$

initiator:
$$- \frac{dI_n}{dt} = 2k_{d_n} I_n \qquad (A-19)$$

Heat Balance

A steady-state heat balance for a plug flow reactor with no radial temperature gradients is given by:

$$\rho C_p \frac{dT}{dZ\tau} = \Delta H R_p - \frac{4h}{D}(T - T_J) - \frac{T}{\rho}\left(\frac{d\rho}{dT}\right)_p \frac{dP}{dZ}\frac{1}{\tau} \qquad (A-20)$$

where

T = temperature
ρ = density
ΔH = heat of reaction
h = heat-transfer coefficient
D = reactor diameter
C_p = heat capacity
R_p = polymerization rate
T_J = reactor jacket temperature
P = pressure
τ = reactor residence time
Z = dimensionless axial distance

and average molecular weight. In particular, the kinetic rate constants specific to high pressure ethylene polymerization were used in the computer-study.

There exists an optimum jacket temperature for maximizing conversion at a given average molecular weight product. The study further suggests that an unstable operating region exists where wide conversion fluctuations result from attempts to increase the reactor conversion by minor adjustments in initiator amount or jacket temperature.

The initiator is the most important reactor parameter in the polymer process. The initiator type affects the molecular weight and conversion limits in a reactor of fixed size and the molecular weight distribution of the material at a given conversion level. The initiator type dictates the initiator amount for a given conversion, the operating temperature range and sensitivity of the reactor to an unstable condition.

Optimized molecular weight-conversion relationship is related to the system heat transfer coefficient. The degree of conversion improvement from improved heat transfer depends on the average molecular weights of polymer being produced for a given initiator system.

Acknowledgements

The authors gratefully acknowledge Union Carbide Corporation for permission to publish this paper.

Literature Cited

1   Lee, K. H. and Marano, J. P., Jr., Paper No. 67d, _AIChE Annual Meeting_, November, 1977, New York.

2   Szabo, J., Luft, G. and Steiner, R., _Chemie-Ing.-Techn._, _41_, 1007 (1969).

3   Ehrlich, P. and Mortimer, G. A., _Adv. Polymer Sci._, 1, 386 (1970).

4   Graessley, W. W., _AIChE-I. Chem. E. Symposium Series, No. 3_, London, 1965.

5   Ray, W. H., Canadian Journal of Chemical Engineering, _45_, 356 (1967).

6   Marano, J. P., Jr., A Seminar Presented at Northwestern University, Evanston, Illinois (February, 1974).

7   Gilles, E. D. and Schuchmann, H., _Chemie-Ing.-Techn._, _38_, 1278 (1966).

8   Ehrlich, P., et al, AIChE Journal, 22, 463, May, 1976.

9   Weale, K. E., Chemical Reactions at High Pressure, Willmer
     Brothers, Ltd., Berkenhead, Great Britain, 217 (1967).

10   Bevington, C., Jr., et al., J. of Polymer Science, 14, 463,
     May, 1976.

11   Bamford, C. H., et al., The Kinetics of Vinyl Polymerization
     by Radical Mechanisms, Butterworths Scientific Pub., London
     (1958).

12   Hammond, G. S., et al., J. Am. Chem. Soc., 77, 3244, (1955).

RECEIVED January 29, 1979.

Molecular Weight Distribution Control in Continuous-Flow Reactors

An Experimental Study Using Feed Perturbations for a Free-Radically Initiated Homogeneous Polymerization in a Continuous-Flow Stirred-Tank Reactor

G. R. MEIRA[1], A. F. JOHNSON[2], and J. RAMSAY

Postgraduate Schools of Polymer Science and Control Engineering,
University of Bradford, Bradford, BD7 1DP, West Yorkshire, England

Our ultimate objective is to produce automatically with laboratory-scale reactors polymers with pre-defined molecular characteristics in reasonable amounts for test purposes. Whatever control is exercised over the chemistry of a polymerization to introduce novel structural features into polymer chains, the final molecular weight distribution (MWD) of the product is always of importance (1,2); hence attention has been given to this subject.

In this short initial communication we wish to describe a general purpose continuous-flow stirred-tank reactor (CSTR) system which incorporates a digital computer for supervisory control purposes and which has been constructed for use with radical and other polymerization processes. The performance of the system has been tested by attempting to control the MWD of the product from free-radically initiated solution polymerizations of methyl methacrylate (MMA) using oscillatory feed-forward control strategies for the reagent feeds. This reaction has been selected for study because of the ease of experimentation which it affords and because the theoretical aspects of the control of MWD in radical polymerizations has attracted much attention in the scientific literature.

A high pressure gel permeation chromatograph (GPC) has been used to monitor the performance of the reactor. A novel aspect of the GPC is that, it too, has been put on-line to the process control computer and both data collection and analysis have been made automatic while giving the operator full interactive facilities.

Molecular Weight Distribution.
    (i) Unperturbed Polymerization Reactors.

There is a great deal of information available on the MWD to be expected in the product from a free-radically initiated chain

[1] Current address: Gas del Estado, Buenos Aires, Argentina.
[2] To whom correspondence should be addressed.

polymerization reaction in various types of idealized reactors (3,4,5) and these data are summarized in Table I. In this table the distribution of chain sizes is described in terms of polydispersity D_n which is defined as

$$D_n = \bar{\mu}_w/\bar{\mu}_n = \bar{M}_w/\bar{M}_n$$

where $\bar{\mu}_w$ and $\bar{\mu}_n$ are the weight average and number average degree of polymerization respectively and $\bar{M}_w$ and $\bar{M}_n$ are the corresponding weight and number average molecular weights.
     There are many interesting reports in the literature where computer simulations have been used to examine not only idealized cases but have also been used in an attempt to explain segregation and viscosity effect in unperturbed polymerization reactors (6). Some experimental work has been reported (7,8). It is obvious, however, that although there is some change in the MWD with conversion in the batch and tubular reactor cases and that broadening of the MWD occurs as a result of imperfect mixing, there is no effective means available for controlling the MWD of the polymer from unperturbed or steady-state reactors.

     (ii) Perturbed Polymerization Reactors. Continuous chemical reactors are usually designed to operate in a steady-state mode. There is evidence which suggests that reactor performance can be improved for processes other than polymerization with forced oscillations in the control variables (9). Bailey (10) has recently reviewed periodic phenomena in chemical reactors and considers the cases of both autonomous and forced oscillations.
     There is less information available in the scientific literature on the influence of forced oscillations in the control variables in polymerization reactions. A decade ago two independent theoretical studies appeared which considered the effect of periodic operation on a free radically initiated chain reaction in a well mixed isothermal reactor. Ray (11) examined a reaction mechanism with and without chain transfer to monomer. The inlet monomer concentration was varied sinusoidally to determine the effect of these changes on D_n^*, the time-averaged polydispersity, when compared with the steady-state case. For the unsteady state CSTR, the pseudo steady-state assumption for active centres was used to simplify computations. In both of the mechanisms considered, D_n^* increases with respect to the steady-state value (for constant conversion and number average chain length $\bar{\mu}_n$) as the frequency of the oscillation in the monomer feed concentration is decreased. The maximum deviation in D_n^* thus occurs as $\omega \to 0$. However, it was predicted that the value of D_n^* could only be increased by 10-32% with respect to the steady state depending on reaction mechanism and the amplitude of the oscillating feed. Laurence and Vasudevan (12) considered a reaction with combination termination and no chain transfer.

Table I.

Influence of Reactor Type on the MWD of the Product of a Free

Radical Polymerization Mechanism.

Reactor	
Homogeneous Batch Reactor.	(i) Low Monomer Conversion (theoretical considerations valid when [M], [I] and $\overline{M}_n$ are held essentially constant) (a) Termination by transfer or disproportionation only: The instantaneous weight chain length distribution becomes the 'most probable' i.e. $D_n \to 2$ for high polymer (b) Termination by combination only: $D_n \to 1.5$ for high molecular weight polymer (ii) High Monomer Conversion: Usually instantaneous M_n tends to fall with conversion increasing D_n. D_n's in the range 2-5 are common but can be much larger.
Homogeneous Steady-State CSTR	Where the mean lifetimes of the growing chains are short, narrower MWD's are produced than in a batch or plug flow reactor but the minimum D_n is 1.5 or 2.0 according to the mechanism of termination. D_n independent of $\overline{\mu}_n$ and $\overline{\mu}_w$.
Segregated Steady-State CSTR	Segregation has little effect but tends to broaden the distribution. As for the case of a batch reactor D_n becomes dependent on $\overline{\mu}_n$ and $\overline{\mu}_w$.
Tubular or Plug flow Reactor.	Under ideal conditions similar to homogeneous batch reactor case.

The inlet concentration of monomer and initiator were each separately varied in a very slow sinusoidal manner. The D_n^* was again predicted to increase in comparison with the non-perturbed case, but they concluded that different results might be observed with regard to the magnitude and direction of the change in the polydispersity under non-isothermal conditions.

Yu ([13]) simulated a periodically operated CSTR for the thermal polymerization of styrene and found the MWD to increase at low frequencies but all effects were damped out at higher frequencies because of the limited heat transfer which occurs relative to the thermal capacity of industrial scale reactors.

Bhawe ([14]) has simulated the periodic operation of a photochemically induced free-radical polymerization which has both monomer and solvent transfer steps and a recombination termination reaction. An increase of 50% in the value of D_n^* was observed over and above the expected value of 2.0. An interesting feature of this work is that when very short period oscillations were employed, virtually time-invariant products were predicted.

The most comprehensive simulation of a free radical polymerization process in a CSTR is that of Konopnicki and Kuester ([15]). For a mechanism which includes transfer to both monomer and solvent as well as termination by combination and disproportionation they examined the influence of non-isothermal operation, viscosity effects as well as induced sinuoidal and square-wave forcing functions on initiator feed and jacket temperature on the MWD of the polymer produced.

One of the few attempts to examine a polymerization reactor in periodic operation experimentally is the work of Spitz, Laurence and Chappelear ([16])who reported the influence of periodicity in the initiator feed to the bulk polymerization of styrene in a CSTR. To induce periodicity the initiator feed was pulsed on-and-off and the reactor output compared with steady-state operation with the same time-averaged initiator input. The objective was to broaden the MWD by forcing initiator concentrations to change with periods long enough to allow marked changes in the reaction environment and short enough to use the reactor as its own blender to dampen the oscillations.

Experimental.

Control Policy. The control variables which, if perturbed, are most likely to influence the MWD of the product of a free radically initiated solution polymerization carried out in a well mixed CSTR are:

(i)   Monomer flow rate, initiator flow rate, or both.
(ii) Monomer concentration, initiator concentration, or both.
(iii) Transfer agent flow rate or concentration.
(iv) Reactor temperature.

In this work only simultaneous perturbations in the monomer and initiator flow rates will be considered.

Although a dynamic mathematical model of the polymerization system has been developed (17) it is not capable of providing the necessary operating policies for the reactor in order to preselect the time-averaged MWD in the product. Hence the flow policies for the reagents were selected empirically and for experimental convenience.

For maximum effectiveness the periods of the oscillation were chosen so as to be relatively long with respect to the hold-up time of the reactor (see Figure 1). A control policy was selected so that the following also pertained.

(i) The time-averaged molecular weight average $\overline{M}^*$ was controlled through the ratio f_M^S / f_I^S where

f_M^S = time-averaged monomer solution flow rate in oscillatory steady-state.

f_I^S = time-averaged initiator solution flow rate in oscillatory steady-state

by making the generally accepted assumption that the instantaneous average molecular weight is proportional to the ratio $[M]/[I]^{0.5}$. Within a certain range it is possible to maintain a constant hold-up time in the reactor, i.e. $f_M + f_I$ = Constant.

(ii) The time-averaged dispersity D_n^* was controlled by oscillating the monomer and initiator flow rates f_M and f_I respectively around their steady-state values. It was assumed that under these conditions, D_n^* would increase while M_n^* would remain essentially constant. Furthermore, for a constant M_n^* it was assumed that D_n^* could be varied between the following limits

(a) a minimum when operating in the steady-flow steady-state with a value between 1.5 and 2.0.
(b) a maximum when applying low-frequency, large amplitude square waves with f_M and f_I being in opposition of phase.

In order to obtain any intermediate values of D_n^* there are various possible approaches which may be adopted which involve changing amplitudes, frequencies and phase angle of the forcing functions. This should be noted but the possibilities will not be considered in any greater detail.

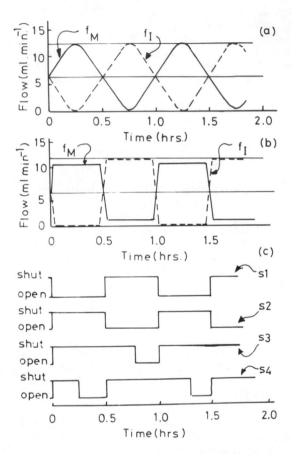

Figure 1.   Forcing functions for monomer (f_M) and initiator (f_I) feeds: (a) sinusoidal; (b) square-wave; (c) reception vessel valve operating sequences which are synchronized with the feed policies (see Figure 2 for the location of the valves S1–S4).

Polymerization Rig.

(i) Reactor System. A schematic representation of the reactor is shown in Figure 2. The stainless steel reactor has a capacity of 300 ml, is fitted with baffles and has a variable speed paddle stirrer. The vessel is heated by means of an oil jacket supplied from a thermostatted tank. The reactor temperature is monitored by means of a thermocouple. The computer controllable pumps (P1 and P2) deliver the monomer and initiator solutions from the stainless steel storage vessels (T1 and T2). The reactor output is taken to a reception vessel system consisting of two 750 ml. glass vessels and four solenoid valves (S1 - S4). The valves were operated by the computer and their on-off positions were synchronised in such a way that it was possible to collect the polymer produced in successive periods of an oscillation when oscilliatory forcing functions were used for the feed pumps. The operation of the valves is shown diagramatically in Figure 1c. This collection method has the advantage over a single holding-tank in that after every oscillation period a representative sample of the product is readily available for analysis. Furthermore, if a new set of quasi steady-state conditions is needed for the control variables, change can be made more rapidly without accumulating off-specification material. Also, the combined volume of the holding-tanks is less than that of a single tank which would give a good 'smoothing' effect.

(ii) Polymer Analysis. A high speed liquid chromatograph (Waters Associates Ltd., ALC/GPC 244) fitted with a U6K injection valve and an appropriate set of μ-Styragel columns was used for the measurement of MWD. This instrument has been linked to an Argus 700 process computer and data analysis made automatic and fast, particularly in comparison with other reported computer aided methods of GPC data analysis (18-27). After a sample injection the computer stores the polymer peak as it emerges, automatically decides when the polymer has completely eluted and immediately afterwards performs the calculations and displays the results on appropriate peripherals without further operator intervention. Conventional calibration and calculation methods have been used (28, 29, 30, 31) and in the work reported here no correction has been made for instrument broadening. The software provides for full interactive facilities and the system has been described elsewhere (32).

(iii) Computer System. A Ferranti Argus 700 E process computer with 64K, 16-bit word core store has been used in this work. The peripherals include a 5 megabyte disk and cassette memory unit, a VDU, teletype and fast line printer. The input and output interface units include 12-bit ADC's, 8-bit DAC's and noise filters.

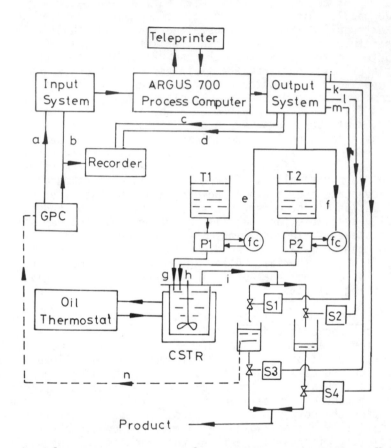

Figure 2.   Schematic representation of the reactor system: computer-controlled pumps (P1, P2); pump controllers (fc); reactor (CSTR); reception vessel valves (S1–S4); monomer and initiator storage vessels (T1, T2). (a) Digital input from GPC injection valve; (b) analogue output from GPC: (c, d) digital outputs to recorder chart drive and event marker; (e, f) analogue outputs for pump set-point adjustment; (g, h) reactor feeds; (i) reactor output; (j–m) digital outputs to reception system valves; (n) manual sampling of products by GPC.

All the computer software has been written in CORAL 66 (33). Three major programs are run in parallel; SIN1A5 (4520 words), operated the control policies for the pumps; GPC1A5 (8095 words), performs the GPC data acquisition and reduction functions; TEL1A5 (3915 words) provides the necessary operator-computer interaction facilities. The control of the timing of any of the three programs mentioned was achieved using standard manufacturers' software (03TAT2)

(iv) Reagents and Reaction Conditions. α,α'-Azobisisobutyronitrile (Koch-Light Laboratories Ltd.) and toluene (May and Baker Ltd.) were used as supplied. Commercial grade methyl methacrylate (Koch-Light Laboratories Ltd.) was vacuum distilled prior to use. Conventional laboratory techniques were used to prepare all monomer and initiator solutions. Polymerizations were carried out under oxygen-free nitrogen and the polymers precipitated from solution with an excess of methanol and vacuum-oven dried. Preliminary experiments were carried out to access the optimum operating conditions, e.g. reactor stirring-rate ($\sim$2000 r.p.m.), monomer concentration ($\sim$ 50% V/V), initiator concentration ($\sim$ 1% W/V) and temperature ($\sim$ 80°C).

Typical Periodic Operation Experiments and Results.

Two typical experiments are described: In the first, sinusoidal forcing functions are used for monomer and initiator feeds to the reactor; the second experiment is similar except that square-wave forcing functions are used. These forcing functions are shown schematically in Figure 1(a,b).
For each experiment the procedure was as follows:

(a) The maximum delivery rate of each pump was set at 11.4 ml/min so that equal set points gave equal flow rates.
(b) With the pump controller in the manual mode the set points were adjusted to 50%, i.e. $f_M^S = f_I^S = 5.7$ ml/min and $f_M^S + f_I^S = 11.4$ ml/min which gave a hold-up time of 25.6 mins. Polymerization was allowed to proceed and the steady-state products were analysed by after 5 or more hold-up times.
(c) Periodic operation of the reactor was introduced by switching the flow controllers to computer control after steady-state had been established with constant pump speeds. After 4-5 periods of oscillation of 1 hour each permanent oscillatory conditions were attained and analysis was carried out on the lumped product of one or more of the subsequent periods.

It is clear from Figure 1 that in the first experiment (Figure 1a) f_M and f_I were oscillated in opposition of phase with maximum possible amplitude of the pump setting. In the second experiment (Figure 1b) nearly rectangular waves were used for

f_M and f_I also 180° out of phase. In this case the f_M was oscillated between 10% and 90% of maximum flow so that at all times some monomer was being introduced to the reactor. The forcing functions were not perfectly rectangular because of the limitations of the controller, which would not respond instantaneously but they did closely approximate to the ideal.

In both experiments the monomer stock solution concentration was 50% V/V (4.63 moles 1^{-1}) and the initiator solution concentrations 0.041 moles 1^{-1} and 0.044 moles 1^{-1} in the sinusoidal and square-wave cases respectively. Reaction temperature was 80°C.

The results are summarized in Table II and the GPC traces are shown in Figure 3.

Table II

The $\overline{M}_n$, $\overline{M}_w$, D_n and conversions observed for the free-radically initiated polymerization of MMA produced in a periodically operated CSTR

Forcing Function	Condition	Curve in Figure 3	$\overline{M}_n$	$\overline{M}_w$	D_n	% Conversion.
Continuous	Steady-state	1	21000	37200	1.77	22.0
Sinusoidal	Oscillatory Steady-state	2	22160	44400	2.0	19.0
Continuous	Steady-state	3	23400	41300	1.76	18.8
Square-wave	Oscillatory Steady-state	4	27000	50900	1.89	18.7

Discussion. It is apparent from Table II and Figure 3 that even though the reactor has been subjected to very severe oscillatory conditions where the frequency of oscillation was low with respect to the hold-up time and the amplitudes of the functions large, the MWD's of the resulting polymers differ very little from those produced in the steady-flow steady-state. Under periodic operation, the polydispersity is greater by between 15-30% and the wave form of the forcing function has little effect. There is a small increase in both $\overline{M}_n^*$ and $\overline{M}_w^*$ with respect to their steady-state values and the conversion appears to be little affected by the mode of operation of the reactor.

It is interesting to compare these findings with those of Spitz et al (16) even though the experimental methods and forcing functions are different in their case. In both studies the time-averaged polydispersity obtained under periodic operation increased by a maximum of 30% when compared with steady-state values. Also when passing from continuous-flow steady-state conditions to permanent periodic operations only a very small drop in monomer conversion was observed in each case. However, contrary to the findings of Spitz et al, $\overline{M}_w^*$ in this work was always slightly higher than $\overline{M}_w$.

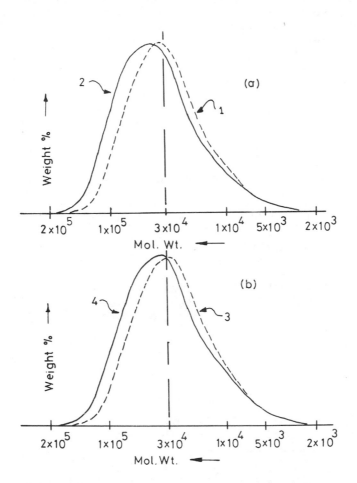

Figure 3. Typical GPC traces—(a) sinusoidal feed: Curve 1, steady-state unperturbed flow; Curve 2, oscillatory steady-state. (b) Square-wave feed: Curve 3, steady-state unperturbed flow; Curve 4, oscillatory steady-state.

The perturbations used in this study are, in general, more severe than those used in the computer models of oscillatory feeds to polymerization CSTR's which have been reported (11, 12, 13, 14, 15) but the influence of the disturbances observed experimentally are probably less than might be expected from the computer models (even though direct comparisons are not really possible). Although our feed-forward control methods could be readily adapted to produce any perturbations we wished in f_M and f_I, a systematic exploration of different forcing functions was not attempted as the likely changes in any of the measured parameters would be within the limits of error to be expected with GPC. It is conceivable that perturbing the temperature of polymerization or the concentration of an effective transfer agent in the reactor might produce greater changes in the MWD's of the products of a radical process.

The influence of changes in these other variables on MWD in a homopolymerization has not yet been tested, but whatever perturbations are introduced to the feed in a radical polymerization in a laboratory-scale CSTR, they are unlikely to introduce dramatic changes in the MWD of the product because of the extremely short life-time of the active propagating chains in relation to the hold-up time of the reactor. This small change in MWD could be advantageous in a radically initiated copolymerization where perturbations in monomer feeds could give control over polymer compositions independent of the MWD. This postulate is being explored currently.

Considerable success has been achieved in controlling the MWD of the products of polymerization where the life-times of propagating centres are long. These studies and others using computer controlled reactors will be reported elsewhere.

Symbols.

D_n = Polydispersity.

D_n^* = Time-averaged polydispersity.

f_M = Instantaneous monomer feed flow-rate.

f_I = Instantaneous initiator feed flow-rate.

f_M^S = Time-averaged monomer solution flow-rate in oscillatory steady-state.

f_I^S = Time-averaged initiator solution flow-rate in oscillatory steady-state.

$[I]$ = Initiator concentration in stock solution.

$[M]$ = Monomer concentration in stock solution.

$\overline{M}_n$ = Number average molecular weight.

$\overline{M}_n^*$ = Time-averaged $\overline{M}_n$

$\overline{M}_w$ = Weight average molecular weight.

$\overline{M}_w^* $ = Time-averaged $\overline{M}_w$

$\overline{\mu}_n$ = Number average degree of polymerization.

$\overline{\mu}_w$ = Weight average degree of polymerization.

Acknowledgment.
   The authors wish to thank the Science Research Council for financial support (Grant No. B/R 82506)

Literature Cited.
1.   Frisch, H.L., 'Physical Chemistry': Enriching Topics on Colloid and Surface Science', Chapter 10, Theorax, La Jolla Cal., 1975.
2.   Martin, J.R., Johnson, J.F. and Cooper, A.R., J. Macromol Sci., Revs. Macromol. Chem., (1972), C8 1, 57.
3.   Flory, P.J., "Principles of Polymer Chemistry', Cornell University Press, Ithica, NY, 1953.
4.   Peebles (Jr), L.H., 'Polymer Reviews', Volume 18, Eds. Mark, H.F., and Immergut, E.H., Interscience, NY, 1953.
5.   Bamford, C.H., Barn, W.G., Jenkins, A.D. and Onyon, P.F., 'The Kinetics of Vinyl Polymerization by Radical Mechanisms'. Butterworths, London, 1958.
6.   Fan, L.T., and Shastry, J.S., J. Polymer Sci., Part D. Macro-molecular Reviews, (1973), 7, 155.
7.   Hamielec, A.E., Hodgkins, J.W. and Tebbins T., A.I.Ch.E. J., (1967), 13, 1087.
8.   Duerksen, J.H., Hamielec, A.E. and Hodgins, J.W., J. Polym. Sci., (1968), C24, 155.
9.   Baccaro, G.P., Gaitonde, J.M. and Douglas, A.I.Ch.E.J. (1970) 16, 249.
10. Bailey, J.E., 'Chemical Reactor Theory', Chapter 12, Eds., Lapidus, L. and Amundson, N.R., Prentice Hall, NY, 1977.
11. Ray, W.H., Ind. Eng. Chem. Process Design Develop., (1968) 7, 442.
12. Laurence, R.L. and Vasudevan, G., Ind. Eng. Chem. Process Design Develop. (1968), 7, 427.
13. Yu. F.C.L., 'Periodic Operation of a Non-Isothermal Polymerization Reactor', M.Sc. Thesis, University of Massachusetts, 1969.
14. Bhawe, M.N., 'A Study on the Application of Periodic Operation for some Chemical Engineering Problems', Ph.D. Thesis, Northwestern University, 1972.
15. Konopnicki, D. and Kuester, J.L., J. Macromol. Sci., Chem., (1974) A8(5) 887.
16. Spitz, J.J. Laurence, R.L. and Chapelear, D.C. 'ACS Symposium Series No. 160, (1976), 72, 86.
17. Bourekas, N., Hodgson, W.G., Johnson, A.F. and Ramsay, J., IUPAC, Macro Madrid, (1974), 1, 27.

18. Gregges, A.R., Bowden, B.F., Barrall, E.M. and
     Horikawa, T.T., Separation Science, (1970), 5, (6), 731.
19. Moore, L.D. and Overton, J.R., J. Chromatography, (1971),
     55, 137.
20. McGraw, J., Sater, V.E. and Kuester, J.L., Decuscope (USA)
     (1973), 12, (2), 2.
21. Hamielec, A.E., Walther, G. and Wright, J.D., 'Advances in
     Chemistry Series'. No. 125, Ed. Ezrin, M., 1973.
22. Maclean, N., American Laboratory, (1974), 16 (10), 63.
23. Ouano, A.C., Horne, D.L. and Gregges, A.R.,
     J. Polym. Sci. Chem. Ed., (1974), 12, 307.
24. Horlzgen, H.J., Chem. Anlagen Verfahren, (1974), (4), 111.
25. Bly. D.D., Du Pont Innovation, (1974), 5 (2), 16.
26. Lesec, J. and Quivoron, Analusis, (1976), 4 (10), 456.
27. Ouano, A.C., J. Chromatography. (1976), 118, 303.
28. Cazes, J., J. Chem. Educ., (1970), 461 and 505.
29. Johnson, J.F. and Porter, R.S., 'Progress in Polymer
     Science', Ed. Jenkins, A.D., Pergamon, NY, 1970.
30. Bly, D.D., 'Physical Methods in Macromolecular Chemistry',
     Volume 2, Ed. Carroll, B., Marcel Dekker Inc., NY, 1972.
31. Evans, J.M., R.A.P.R.A. Bulletin, Nov. 1972.
32. Meira, G.R., Johnson, A.F. and Ramsay, J. (in the press)
33. Webb, J.T., 'CORAL 66 Programming', NCC Publications,
     Oxford, 1978.

RECEIVED March 12, 1979.

A Review of Mechanistic Considerations and Process Design Parameters for Precipitation Polymerization

M. R. JUBA

Research Laboratories, Eastman Kodak Company, Rochester, NY 14650

Precipitation polymerizations generally consist of two phases: the diluent phase and the solid polymer particles.

The diluent is a solvent for the monomer and initiator and a nonsolvent for the polymer.

The polymer particles are not stabilized and tend to agglomerate to form a polymer paste or slurry. In addition, the polymerization rate is independent of the number of particles (1).

Some typical examples of precipitation polymerizations are:

Monomer	Diluent	
methyl methacrylate	cyclohexane	(2)
styrene	methanol	(3)
acrylonitrile	bulk	(4)
vinylidene chloride	bulk	(5)
vinyl chloride	bulk	(6)

Since these precipitation polymerizations produce polymeric solids at very high polymerization rates and very high purity (i.e., free from emulsifiers, suspending agents, etc.), their popularity as manufacturing processes is increasing. This creates some interesting challenges for the process design engineer who is searching for a relationship among the reaction parameters and the physical variables of the reaction.

These relationships are generally determined empirically, because of the complex kinetics of the precipitation polymerization process and the large variations from one reaction system to another. Nevertheless, a review of the literature presents useful guidelines for process design experiments.

Particle Formation. Electron microscopy and optical microscopy are the diagnostic tools most often used to study particle
formation and growth in precipitation polymerizations (7,8).
However, in typical polymerizations of this type, the particle
formation is normally completed in a few seconds or tens of
seconds after the start of the reaction (9), and the physical
processes which are involved are difficult to measure in a real
time manner. As a result, the actual particle formation mechanism is open to a variety of interpretations and the results
could fit more than one theoretical model. Barrett and Thomas
(10) have presented an excellent review of the four physical
processes involved in the particle formation:
    oligomer growth in the diluent
    oligomer precipitation to form particle nuclei
    capture of oligomers by particle nuclei, and
    coalescence or agglomeration of primary particles.

The first process begins as initiator decomposes in the
diluent phase and polymerizes the monomer to form oligomers
which precipitate from solution upon reaching a critical molecular weight.
    This critical molecular weight increases with the solubility
of the polymer and is low enough so that all the oligomers are
captured or nucleate particles before their radicals are terminated. As a result, nearly all polymerization takes place in
the particles and the polymer concentration in the diluent phase
is low.
    The polymer solubility can be estimated using solubility
parameters (11) and the value of the critical oligomer molecular
weight can be estimated from the Flory-Huggins theory of polymer
solutions (12), but the optimum diluent is still usually chosen
empirically.
    Barrett and Thomas (10) listed the following effects of
increasing the diluent's solvency:
    retards the onset of particle formation,
    increases the duration of particle formation, and
    produces fewer, larger particles with a broader particle
     size distribution.

Once the oligomers have formed, two mechanisms, self-
nucleation and aggregate nucleation, are used to describe
particle nucleation.
    In self-nucleation, the extended oligomer chain collapses
upon itself to nucleate a particle.
    In aggregate nucleation the oligomers reversibly associate
with each other until the aggregate reaches a critical size
above which it is thermodynamically stable and continues to
grow.
    The aggregation of oligomers requires a lower average
degree of polymerization for nuclei formation than the self-
nucleation model where a larger individual chain is required.

Therefore, aggregation is considered the primary mode of particle nucleation in most systems.

After the particle nuclei form, they capture the oligomers growing in the diluent phase and essentially no new particle nuclei are formed. Two models have been employed to explain this capture.

The first is diffusion capture. This theory was originally proposed by Fitch and Tsai (13) for the aqueous polymerization of methyl methacrylate. According to this theory, any oligomer which diffuses to an existing particle before it has attained the critical size for nucleation is irreversibly captured. The rate of nucleation is equal to the rate of initiation minus the rate of capture. The rate of capture is proportional to both the surface area and the number of particles.

In the equilibirum capture model, on the other hand, there is a dynamic equilibrium between the growing oligomers and the surface of the particles as well as the possibility of some interchange with the interior of the particles.

Although both of these models provide a reasonable description of the precipitation polymerization process, they do not illustrate the relationship between the reactor variables and the polymer particle properties.

Perhaps the only process where such correlations have been published is the bulk polymerization of vinyl chloride as reported by Ray, Jain and Salovey (14).

The polymerization occurs in four stages.

Bulk PVC Process

Nucleation of the primary particle population
Flocculation of particles and capture of oligomers to a
  point of constant particle population
Polymerization until the separate monomer phase is consumed
Polymerization of absorbed monomer in the polymer particles

The commercial process as described by J. Chatelain (15) consists of two stages as shown in Figure 1.

In the prepolymerizer, the polymer particles are formed and polymerized to 7-8 wt. % conversion before being transferred to the autoclave where the particles are polymerized to a solid powder at about 88% conversion.

The final polymer particles have a narrow particle size distribution, Figure 2 (15), and the mean particle size is a strong function of the agitation in the prepolymerizer, Figures 3 and 4 (16).

In addition, several patents, discuss the effects of various additives on the particle size of the final product.

Ray, Jain and Salovey (14) modeled these phenomena using the kinetic constant of coalescence as their major parameter.

This constant was a function of particle size, agitation rate, and the surface properties of the particles, and its functional form suggested that the probability of coalescence was proportional to the surface area per unit volume of the

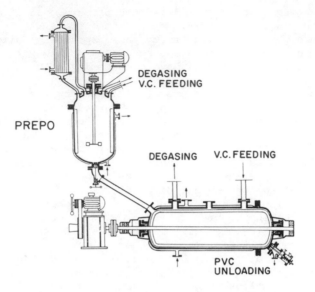

British Polymer Journal

Figure 1.   Reactor system for bulk PVC (15)

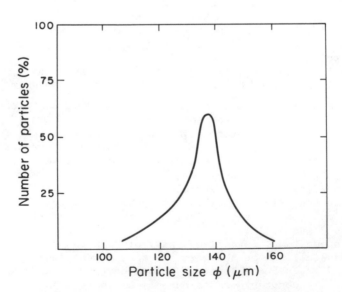

British Polymer Journal

Figure 2.   Particle size distribution for bulk PVC (15)

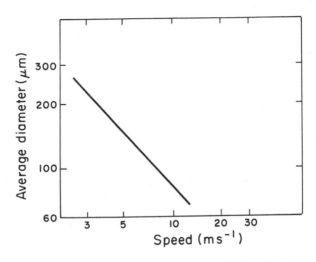

Society of Plastics Engineers Journal

Figure 3. Effect of prepolymerizer agitator speed on the mean particle size of bulk PVC (16)

Effects of agitator size and speed
on bead size (microns).

Turbine diameter

Speed rpm	10 in. 254 mm	12 in. 305 mm	14 in. 356 mm
230	200	180	160
300	170	150	120
350	140	130	90

Society of Plastics Engineers Journal

Figure 4. Effects of prepolymerizer agitator size and speed on the mean particle size of bulk PVC (16)

coalescing particles.

The coalescence constant required experimental correlation with the agitation rate and the surface free energies of the polymer particles.

While the model was in general agreement with the limited experimental data published on bulk PVC particle size distribution, there is still no generally applicable theory describing particle growth and flocculation in the presences of mechanical agitation for precipitation polymerizations.

Polymerization Rate and Radical Occlusion. In modeling the rate of precipitation polymerization, the reaction can occur at three different loci: in the diluent, at the surface of the particles, or in the interior of the particles.

The predominant mode of polymerization is in the interior of the particles and this leads to a reduction of macroradical mobility, usually referred to as radical occlusion, and a marked autoacceleration of the polymerization rate.

Some typical examples of this autoacceleration are: (Figure 5) Norrish and Smith (2) polymerized methyl methacrylate in bulk and in the presence of various precipitants and measured the polymerization rates dilatometrically. They determined that autoacceleration of the precipitation polymerizations was larger than that observed for the Trommsdorf effect in bulk polymerization.

Similarly, Garcia-Rubio and Hamielec (17) conducted bulk polymerizations of acrylonitrile at various temperatures and initiator levels in glass ampoules. Their plots of the rate of polymerization as a function of conversion are typical of the extensive radical occlusion in this very glassy polymer.

Finally, similar autoacceleration in the polymerization rate was reported by Crosato-Arnaldi, Gasparini and Talamini (18) for the bulk polymerization of vinyl chloride.

In addition, Bamford, Jenkins and coworkers (19) previously reported on the behavior of occluded radicals in the heterogeneous polymerizations of acrylonitrile, methyl acrylate, methyl methacrylate and vinylidene chloride. From their electron spin resonance studies, they concluded that the degree of occlusion was:

1)   Largest for glassy polymers like acrylonitrile which are not highly swollen by monomer. (Living macroradicals can be obtained in heterogeneous acrylonitrile polymerization.)

2)   Decreasing as the reaction temperature exceeded the glass transition temperature of the monomer polymer mixture.

3)   Decreasing with swelling of the polymer matrix.

4)   Decreasing with chain transfer to monomer or transfer agents. (This process is not as limited by diffusion as macroradical-macroradical termination reactions.)

Since the extent of radical occlusion varies from one precipitation polymerization to the next, it is nearly impossible to develop a generalized polymerization rate equation. As a result, rate expressions are most often determined from experi-

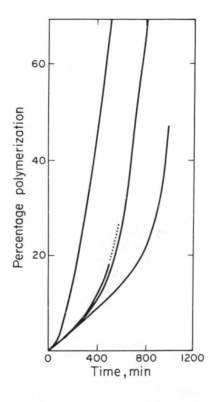

Nature

Figure 5. Effect of autoacceleration on the precipitation polymerization of methyl methacrylate (2). The curves, from left to right, are for the diluents: cyclohexane; t-butylsterate; heptane; and bulk.

mental data. Since it is very difficult to reproduce dilatomet-
ric or gravimetric rate measurements of precipitation polymerizations, particularly at the high conversions of interest to the
process design engineer, microcalorimetric techniques, such as
the one discussed by Barrett and Thomas (20) which uses a DSC
cell, are gaining in popularity.

Copolymerization. The solid phase of the precipitation
polymerization also influences copolymer composition, since differential monomer adsorption on the polymer particles considerably
modifies the effective reactivity ratios of the comonomers. This
problem has been discussed by several authors (22,23,24,25,26).
Two particularly interesting examples are:
1. The heterogeneous copolymerization of styrene and acryloni-
    trile in various diluents as reported by Riess and Desvalois
    (22). Although the copolymer composition in these studies
    was not strongly influenced by the diluent choice, the pre-
    ferential adsorption of acrylonitrile monomer onto the
    polymer particles shifted the azeotropic copolymerization
    point from the 38 mole % acrylonitrile observed in solution
    to 55 mole % acrylonitrile.
2. Myagchenkov and coworkers (23) reported that no reasonable
    reactivity ratios could be chosen for the heterogeneous
    copolymerization of acrylamide and maleic acid in dioxane.
    Barrett and Thomas (10) proposed that these effects of dif-
ferential monomer adsorption could be modeled by correcting
homogeneous solution copolymerization reactivity ratios with
the monomer's partition coefficient between the particles and
the diluent. The partition coefficient is measured by static
equilibrium experiments. Barrett's suggested equations are:

$$[M_1/M_2]_p = F \ [M_1/M_2]_d$$

$$r_1' = Fr_1 \qquad r_2' = r_{2/F}$$

Where:

F = The monomer's partition coefficient between the diluent
    and the particles

$[M_1/M_2]_d$ = Ratio of monomer one to monomer two in the diluent

$[M_1/M_2]_p$ = Ratio of monomer one to monomer two in the particles

r_1 = Reactivity ratio of monomer one in a homogeneous solu-
    tion polymerization

r_2 = Reactivity ratio of monomer two in a homogeneous solu-
    tion polymerization

r_1' = Reactivity ratio of monomer one in a precipitation
    polymerization

r_2' = Reactivity ratio of monomer two in a precipitation
    polymerization

However, this approach does not account for:
A. Changes of the partition coefficient with monomer consumption
   in the diluent phase.
B. Effects of starvation feeding in semicontinuous or continuous
   polymerization systems.
C. The possibility of monomer gradients in the glassy particles.

Seymour and coworkers (27,28,29,30) actually used these
composition gradients to prepare block copolymers by swelling
particles containing occluded (i.e., living) macroradicals with
a second monomer. Such block copolymers were prepared from
occluded vinylacetate, methyl methacrylate, and acrylonitrile
macroradicals, and the yield of block copolymers was studied as
a function of the solubility and rate of diffusion of the swelling
monomer in the particles.

Reactor Design Considerations

The literature on the modeling and design of precipitation
polymerization reactors is limited primarily to reactor for the
bulk polymerization of vinyl chloride (31-38), although other
systems have been discussed, particularly in the patent literature (39,40,41).

As is common in most polymer reactor design problems, heat
transfer is one of the major process concerns. For example, if
the heat transfer is primarily through the wall of a jacketed
reactor, the overall heat transfer coefficient is a function of
both the agitator configuration and the degree of swelling of
the particles.

Two patents (41,42) discuss the design of special agitators
to maintain adequate heat transfer in bulk polyvinyl chloride
reactors.

The effect of the particle properties on the overall heat-
transfer coefficient was investigated in our laboratory (43) for
an acrylic precipitation polymerization as shown in Figure 6.
The polymerizations were conducted in a 20-liter stainless steel
reactor with a pitched-blade turbine agitator and four side-wall
baffles. The monomer was polymerized at the same temperature,
initiator and monomer concentration in two different inert
diluents. The data (Figure 6) illustrate the substantial lowering
of the overall heat transfer coefficient for the system with the
more highly swollen particles.

This is an important consideration in the selection of an
optimum polymerization diluent, which is very easily neglected
in laboratory investigations. Also, since little is known about
particle coalescence in the presence of mechanical agitation,
extreme care must be taken in mixing scale-up.

Perhaps the most challenging reactor design problem is the
design of the continuous precipitation polymerization reactor.
Although several patents (45,46,47) are concerned with this problem, the topic has been generally neglected in the reactor-

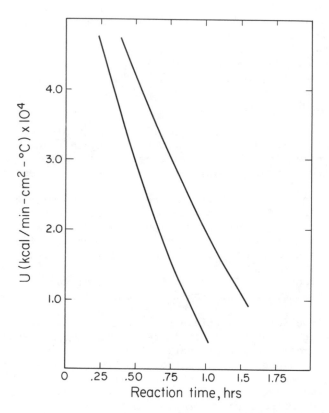

Figure 6.   Reduction of the overall heat transfer coefficient attributable to particle swelling in an acrylic precipitation polymer (43)

design literature.

Friis and Hamielec (48) offered some comments on the continuous reactor design problem suggesting that the dispersed particles have the same residence time distribution as the dispersing
fluid and the system can be modeled as a segregated CSTR reactor.

However, since virtually no published data are available on
the mechanism of continuous particle formation and growth,
particlarly under the influence of mechanical agitation, continuous precipitation polymerization remains a vast engineering
wilderness waiting for experimental and theoretical explorations.

Summary

In summary, there are no generalized models which can be
applied to all precipitation polymerization systems, and careful
experimental investigations are required for successful process
designs.

Most of the studies reported in the literature are based on
batch polymerization systems. For these systems an excellent
foundation has been laid for further studies.

The major unresolved questions for these systems are the
coalescence and flocculation of particles under mechanical agitation and the parameters which influence copolymer composition
drift.

On the other hand, very little is known about the mechanism
of continuous precipitation polymerization. In particular, the
mechanism of continuous particle formation and growth and the
effects of starvation feeding on reaction rate and copolymer
composition are areas of particular interest.

Abstract

The precipitation polymerization literature is reviewed
with particular attention to the influence of particle formation
and growth, autoaccelerating polymerization rates, and copolymer
composition drift on polymer reactor design.

Literature Cited

1. K. E. J. Barrett and H. R. Thomas, Kinetics and Mechanism
    of Dispersion Polymerization, in Dispersion Polymerization
    in Organic Media, K. E. J. Barret, Ed., F. Wiley and Sons,
    London, 1975.
2. R. G. W. Norrish and R. R. Smith, Nature 150 1942 336.
3. M. Hoffmann, Makromol. Chem. 177 (4) 1976 1021-50.
4. C. H. Bamford and A. D. Jenkins, Proceedings of the Royal
    Society, A228 1955 220.
5. C. H. Bamford, W. G. Barb, and A. D. Jenkins, The Kinetics of
    Vinyl Polymerization by Radical Mechanisms, Academic Press,
    New York, 1958, 117.

6. J. Ugelstad, H. Flogstad, T. Hertzberg, and E. Sund, Makromol. Chem. 164 1973 171-181.
7. W. M. Thomas, Advances in Polymer Science 13 1961 401.
8. J. D. Cotman, M. F. Gonzalez, and G. C. Claver, J. Polym. Sci. A-1 5 1967 1137-1164.
9. K. E. J. Barrett and H. R. Thomas, J. Polym. Sci. A-1 7 1969 2621-2650.
10. K. E. J. Barrett and H. R. Thomas, Kinetics and Mechanism of Dispersion Polymerization, op. cit., 144.
11. K. L. Hoy, J. Paint Technol. 42 76 (1970).
12. P. J. Flory, Principles of Polymer Chemistry, Cornell University Press, 1953, 512.
13. R. M. Fitch and C. H. Tsai, in Polymer Colloids (ed. Rim. Fitch) Plenum Press, New York, 1971, 73.
14. W. H. Ray, S. K. Jain, and R. Salovey, J. Appl. Polym. Sci. 19 1975 1297-1315.
15. J. Chatelain, Br. Polym. J. 5 1973 457-465.
16. J. C. Thomas, SPE J., 23 October, 1976 65.
17. L. H. Garcia-Rubio and A. E. Hamielec, Private Communication to M. R. Juba, June 1977.
18. A Crosato-Arnaldi, P. Gasparini, and G. Talamini, Makromol. Chem. 117, 1968 140.
19. C. H. Bamford, A. D. Jenkins, M. C. R. Symons, and M. G. Townsend, J. Polym. Sci. 34 1959 181-198.
20. K. E. J. Barrett and H. R. Thomas, J. Polym. Sci. A-1 op. cit.
21. M. R. Juba and C. J. Smith, unpublished work.
22. G. Riess and M. Desvalois, J. Polym. Sci., Polym. Let. Ed., 15 1977 49-54.
23. V. A. Myagchenkov, V. F. Kurenkov, and S. Ya. Frenkel, Eur. Polym. J. 6 1970 1649-1654.
24. N. N. Slavnitskaya, Yu. D. Semchikov, and A. V. Ryabov, Tr. Khim. Khim, Tekhnol. (2) 1970 90-93.
25. N. N. Slavnitskaya, Yu. D. Semchikov, and A. V. Ryabov, Vysokomol. Soedin. Ser. B 9 (12) 1967 887-890.
26. N. N. Slavnitskaya, Yu. D. Semchikov, and A. V. Ryabov, and D. N. Bort, Vysokomol. Soedin. Ser. A 12 (8) 1970 1756-62.
27. R. B. Seymour, et al. J. Appl. Polym. Sci., Appl. Polym. Sym. 25 1974 69.
28. R. B. Seymour and G. A. Stahl, Advances in Chemistry Series, A.C.S. Washington, D. C. 142 1975 309.
29. R. B. Seymour, et al. Advances in Chemistry Series, A.C.S. Washington, D. C. 129 1973 230.
30. R. B. Seymour and G. A. Stahl, J. Polym. Sci., Polym. Chem. Ed. 14 1976 2452.
31. D. W. Eastman and G. C. Hopkins, Ger. Offen. 2,240,252 February 22, 1973.
32. A. L. Lemper, Ger. Offen. 2,236,428, February 15, 1973.
33. L. Jourdan, Chim. Ind., Genie Chim. 106 (7) 1973 475-7.
34. D. Feldman and M. Macoveanu, Stud. Cercet. Chim. 20 (7) 1972 863-80.

35. J. Chatelain, Nuova Chim. <u>48</u> (8) 1972 85-7.
36. F. Fournel and S. Soussan, Ger. Offen. 2,407,054, September 5, 1974.
37. Hooker Chemical and Plastics Corp., U.S. Patent 4,029,863, July 21, 1975.
38. P. Melacini, L. Patron, and G. Donia, Ger. Offen. 2,428,093, January 16, 1975.
39. E. Scobel, Ger(East) 111,211, February 5, 1975.
40. J. Aleman, Rev. Plast. Mod. <u>32</u> (242) 1976 217-29.
41. P. P. Rathke, U.S. Patent 3,799,917, March 26, 1974.
42. P. P. Rathke, Ger. Offen. 2,307,462, August 30, 1973.
43. M. B. Rivers and M. R. Juba, unpublished work.
44. J. C. Floros, U.S. Patent 3,759,879, May 28, 1971.
45. J. B. Busby and D. A. Hughes, Australian Patent 466,488, October 30, 1975.
46. J. Nelles, et al. Ger(East) 81,721, May 5, 1971.
47. H. Brink Mann, et al. Ger. Offen. 2,141,770, March 8, 1973.
48. N. Friis and A. E. Hamielec, Principles of Polymer Reactor Design, in <u>Polymer Reaction Engineering</u> Course Notes, McMaster University, Hamilton, Ontario, Canada, p.55.

RECEIVED February 21, 1979.

The Anionic Solution Polymerization of Butadiene in a Stirred-Tank Reactor

J. G. MOORE, M. R. WEST, and J. R. BROOKS

Department of Chemical Engineering, University of Leeds, U.K. LS2 9JT

The research programme into n-butyl lithium initiated, anionic polymerization started at Leeds in 1972 and involved the construction of a pilot scale, continuous stirred tank reactor. This was operated isothermally, to obtain data under a typical range of industrial operating conditions.

Mathematical models of the reaction system were developed which enabled prediction of the molecular weight distribution (MWD). Direct and indirect methods were used, but only distributions obtained from moments are described here. Due to the stiffness of the model equations an improved numerical integrator was developed, in order to solve the equations in a reasonable time scale.

It has been possible to obtain a good measure of agreement between the experimental results, and those predicted by even a simple mathematical model of the system, assuming ideal stirred tank behaviour. One typical result is presented here.

Description of the Experimental System.

The experimental investigation used a 3 litre mild steel CSTR designed and constructed within the department of Chemical Engineering at Leeds University and depicted in Fig. 1, which was capable of operation at temperatures up to 423K and pressures up to 9 bar. This was fitted with a single helical ribbon impeller driven at 60 r.p.m., to ensure good mixing of the reactor contents. The reactor could be heated by use of electrical heating tapes wound round its external surface and cooled by a flow of water through an internal coil. The reactor was fully instrumented with respect to process conditions, the instruments being interfaced to a computer system, to allow on-line data acquisition, and eventually control. The reactor pressure was measured by a force balance transducer. Two thermocouples measured the temperature of the reactants at the top and bottom of the reactor. The impeller design required that the thermocouples entered the reactor through its base plate, together with the cooling coil. The reactant volume was measured

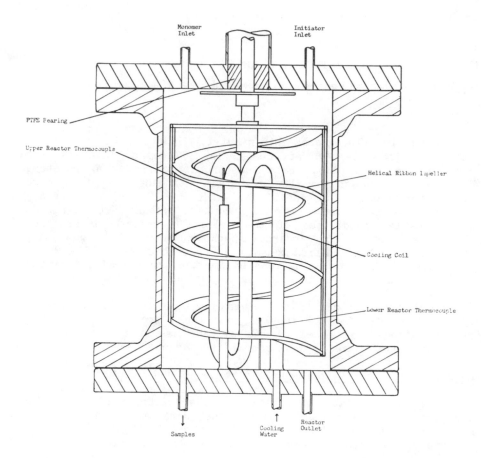

Figure 1.   Three-liter reactor

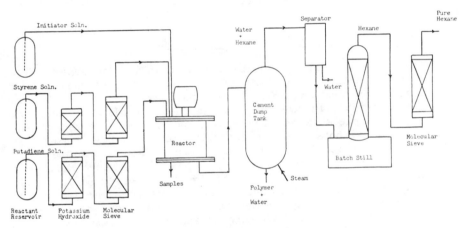

Figure 2. System flowsheet

by a differential pressure transducer, and was controlled from the computer by means of a solenoid valve in the outlet line.

The experimental rig was constructed to minimise the chance of reactant contamination by oxygen and moisture. The feed solutions were made up in storage vessels capable of withstanding 11 bar. High pressure nitrogen was used to drive the solutions into the reactor, eliminating the opportunity for impurities to seep through the packings of any pumps. The monomer feeds passed through towers containing potassium hydroxide, which removed the inhibitor. Then they were driven through drying towers containing molecular sieve type 4A, after which their moisture contents were monitored by a continuous hygrometer. The feed flows were measured using rotameters, fitted with float following devices, which enabled the flowrate to be transmitted to the computer.

The solvent was recovered from the polymer cement by steam stripping, followed by the separation of the organic layer. It was then purified in a small batch still, incorporating a packed column and dried by passage through a bed of molecular sieve, before being returned to the storage vessels.

Operating Conditions and Experimental Methods

The experimental programme was mainly concerned with estimating kinetic parameters from isothermal steady state operation of the reactor. For these runs, the reactor was charged with the reactants, in such proportions that the mixture resulting from their complete conversion approximated the expected steady state, as far as total polymer concentrations was concerned. In order to conserve reactants, the reactor was raised to the operating temperature in batch mode. When this temperature had been attained, continuous flow operation commenced. This was maintained for several residence times. Runs were carried out at 363K, 384K and 393K using monomer feed concentrations of between 5 and 25%. The initiator feed concentration was maintained around .01 mol/l. Residence times varied from 60 to 120 minutes. In practice the reactant volume was maintained at 2.7 l, as this improved the controllability of the system. During experimentation the process conditions were recorded automatically, while the feed solutions, and the reactor contents were sampled approximately every hour during the flow reactor operation phase, and subsequently analysed off-line. The monomer feed streams were analysed by gas-liquid chromatography. The initiator feed was determined by titration using the method of Gilman (1) Samples of the reactor contents were quenched with methanol and the polymer precipitated with acetone. The polymer content was determined gravimetrically and its molecular weight distribution by gel permeation chromatography. The microstructure was determined using infra-red absorption techniques.

The Framework for the Mathematical Modelling

Important features of the modelling work are the means of integration of the model equations and the method of regenerating the dynamic polymer distribution from its moments. The framework provided by this approach makes it possible to produce models with few assumptions about the model behaviour.

The integrator that has been developed is designed for the solution of stiff systems of ordinary differential equations (ODEs) since the differential equations for the higher moments introduce considerable stiffness into the system. The integrator uses Gear's method (2,3,4), an implicit predictor-corrector algorithm. The implementation has been shown to be more efficient than other implementations of Gear's method (3). The integrator can be accessed through several different subroutines which give the user varying degrees of control over the facilities available, the objective being to make the integrator at once easy to use, yet flexible enough for the most demanding user. Such flexibility is important for the solution of the type of model considered here, since development can occur around the model equations rather than around the limitations of the numerical integrator.

Methods for generating distribution from moments have been available since the last century. They were used originally as a method of fitting a distribution curve through poor data, but they are equally well suited for generating a curve directly from the moments.

Two types of curve have been fitted to the moments. The first of these is the Pearson Distribution (5), a curve which is described by the differential equation

$$\frac{dy}{dx} = \frac{x - a}{b_c + b_1 x + b_2 x^2} \tag{1}$$

The second method that has been developed is the Laguere Polynomial, which will be more familiar from its use by Bamford and Tompa (6).

Both methods have advantages. The Laguere Polynomial has the advantage that it can be used to fit almost any curve. The disadvantages are that it can never give the exact distribution, even where one could be given, and unless the shape is characteristic of a Laguere Polynomial, convergence can be slow. The Pearson Distribution has the advantage of giving the exact distribution in a number of cases, and it only requires four moments. However, for distribution curves that are not of the Pearson Type, completely erroneous curves may be generated. A characteristic of both types of curve is that the error is predominantly in the tails of the distribution. Where these methods give good agreement on the distribution curve, confidence can be placed in the result.

The multiple reactor capability allows the modelling of up to five CSTRs connected in any possible configuration. This is achieved by simple mass and energy accounting, with the user supplying the required stream splits. It will be noted that the same model can be used for investigating the effect of poor mixing within a single reactor.

A Model of an Anionic Polymerisation System

The reaction scheme considered in this model is described by

$$\text{Rate of initiation} = k_i' \left[RLi\right]_f^x \left[M\right] \tag{1}$$

$$\text{Rate of propagation} = k_p' \left[RLi\right]_f^y \left[M\right] \tag{2}$$

This complex rate expression can be used to model reactions of the type

$$(RLi)_n \rightleftharpoons nRLi \qquad K_{IA} = \frac{\left[RLi\right]^n}{\left[(RLi)_n\right]} \tag{3}$$

$$RLi + M \longrightarrow R.M.Li \quad k_i \tag{4}$$

The rate of initiation can be expressed as

$$\text{Rate}_i = (k_i K_{IA}^{1/n}/n) \left[RLi\right]_f^{1/n} \left[M\right] \tag{5}$$

$$k_i' = k_i K_{IA}^{1/n}/n \qquad , \qquad x = 1/n \qquad , \tag{6}$$

if it can be assumed that the rate of exchange between associated initiator and 'active' initiator is high relative to the overall rate of initiation. It can also be used as it stands as a complex rate equation for systems where the mechanism is more complex.

The propagation rate expression can be used to describe simple dissociative schemes of the type

$$(PLi)_n \rightleftharpoons nPLi \qquad\qquad K_{PA} \tag{7}$$

$$K_{PA} = \frac{\left[PLi\right]^n}{\left[(PLi)_n\right]}$$

$$RM_n Li + M \longrightarrow RM_{n+1}Li \qquad\qquad k_p \tag{8}$$

Here the rate of propagation can be expressed as

$$\text{Rate}_p = (k_p K_{PA}^{1/n} n) \left[PLi\right]_f^{1/n} \left[M\right] \tag{9}$$

$$k_p' = k_p K_{PA}^{1/n}/n \qquad , \qquad y = 1/n \tag{10}$$

Mass Balance on Initiator

$F_{I,j}I_0$ ——Mass of initiator entering in the inlet stream for reactor j.

$I_{00,j}$ ——Mass of initiator entering reactor j from other reactors.

$I_j F_{R,j}$ ——Mass of initiator leaving reactor j.

$V_j I_j^x M_j k_I'(T_j) = \text{Rate}_{I,j} V_j$ —— Mass of initiator consumed by initiation reaction in reactor j.

$$\frac{d(I_j V_j)}{dt} = F_{I,j}I_0 + I_{00,j} + I_j F_{R,j} - V_j \text{Rate}_{I,j} \qquad (11)$$

Since V_j is constant,

$$\frac{d(I_j)}{dt} = (F_{I,j}I_0 + I_{00,j} - I_j F_{R,j})/V_j - \text{rate}_{I,j} \qquad (12)$$

Mass Balance on Monomer

$F_{M,j}M_0$ ——mass of monomer entering reactor j in the inlet stream.

$M_{00,j}$ ——mass of monomer entering reactor j from other reactors.

$F_{R,j}M_j$ ——mass of monomer leaving reactor j.

$V_j I_j^x M_j k_I'(T_j) = V_j \text{Rate}_{I,j}$ ——mass of monomer consumed in initiation reaction.

$V_j M_j U_0^y k_P'(T_j) = V_j \text{Rate}_{p,j}$ ——mass of monomer consumed in propagation reaction

$$\frac{d(M_j V_j)}{dt} = F_{M,j}M_0 + M_{00,j} - F_{R,j}M_j - V_j \text{Rate}_{I,j} - V_j \text{Rate}_{p,j} \qquad (13)$$

$$\frac{d(M_j)}{dt} = (F_{M,j}M_0 + M_{00,j} - F_{R,j}M_j)/V_j - \text{Rate}_{I,j} - \text{Rate}_{p,j} \qquad (14)$$

Mass Balance on Polymer

$P_{00,n,j}$ ——Polymer of chain length n from other reactors (mass) entering reactor j.

$F_{R,j}P_{n,j}$ ——mass of polymer leaving reactor j of chain length n.

$V_j P_{n,j} M_j k_p(T_j)$—mass of polymer of chain length n destroyed in reactor j.

$V_j M_j P^*_{n-1,j} k_p(T_j)$ —mass of polymer of chain length n created in reactor j.

P   concentration of polymer (total).

P*   concentration of polymer (unassociated).

For n greater than 1:

$$\frac{d}{dt}(P_{n,j} V_j) = P_{00,n,j} - F_{R,j} P_{1,j}$$

$$- V_j P^*_{n,j} M_j k_p(T_j) + V_j P^*_{n-1,j} M_j k_p(T_j) \tag{15}$$

$$\frac{d}{dt}(P_{n,j}) = (P_{00,n,j} - F_{R,j} P_{n,j})/V_j - P^*_{n,j} M_j k_p(T_j)$$

$$+P^*_{n-1,j} M_j k_p(T_j) \tag{16}$$

For n=1

$$\frac{d}{dt}(V_j P_{1,j}) = P_{00,1,j} - F_{R,j} P_{1,j} + V_j Rate_{I,j}$$

$$- V_j P^*_{1,j} M_j k_p(T_j) \tag{17}$$

$$\frac{d}{dt}(P_{1,j}) = (P_{00,1,j} - F_{R,j} P_{1,j})/V_j + Rate_{I,j}$$

$$-V_j P^*_{1,j} M_j k_p(T_j) \tag{18}$$

Polymer Moments

Defining the moments by,

$$U_{0,j} = \sum_n P_{n,j} \tag{19}$$

$$U_{1,j} = \sum_n n P_{n,j} \tag{20}$$

$$U_{1,j} = \sum_n n^1 P_{n,j} \tag{21}$$

$$\frac{dU}{dt}_{1,j} = \sum_n n^1 \frac{dP}{dt}_{n,j} \tag{22}$$

from (16) and (18) one can obtain

$$\frac{dU}{dt}_{1,j} = (P_{00,1,j} - F_{R,j}P_{1,j})/V_j + \text{Rate}_{I,j} - P^*_{1,j}M_j k_p(T_j)$$

$$+ \sum_n (n+1)^1 \left\{ (P_{00,n+1} - F_{R,j}P_{n,j})/V_j + P^*_{n,j}M_j k_p(T_j) \right.$$

$$\left. - P^*_{n+1,j}M_j k_p(T_j) \right\} \tag{23}$$

Rearranging

$$\frac{dU}{dt}_{1,j} = \text{Rate}_{I,j} - (R_{R,j}/V_j)\sum_n n^1 P_{n,j} + (\frac{1}{V})\sum_n P_{00,n,j}$$

$$- \sum_n n^1 (P^*_{n,j}M_j k_p(T_j)) + \sum_n (n+1)^1 (P^*_{n,j}M_j k_p(T_j)) \tag{24}$$

$$= \text{Rate}_{I,j} - (F_{R,j}/V_j)U_{1,j} + U_{00,1,j}/V_j$$

$$- U^*_{1,j}M_j k_p(T_j) + (\sum_{i=00}^{1} (\tfrac{1}{i})U^*_{1-i,j})M_j k_p(T_j) \tag{25}$$

Now assuming that the rate of exchange between associated and unassociated polymer chains is rapid, compared to the rate of propagation, then the distribution of the active polymer, will be equivalent to the total polymer distribution.

I.e.
$$\frac{U^*_{1,j}}{U^*_{0,j}} = \frac{U_{1,j}}{U_{0,j}} \tag{26}$$

Now
$$U^*_{0,j} = yK^y_{PA} U^y_{0,j} \tag{27}$$

So
$$U^*_{1,j} = (U_{1,j}/U_{0,j}) U_{0,j} yK^y_{PA}$$

$$= U_{1,j} U^{y-1}_{0,j} yK_{PA}$$

Thus
$$\frac{dU}{dt}_{1,j} = \text{Rate}_{I,j} - (R_{R,j}/V_j) U_{1,j} + U_{00,1,j}/j$$

$$- \left[U_{1,j} U^{y-1}_{0,j} yK^y_{PA} \right] M_j k_p(T_j)$$

$$+ \left[\sum_{i=0}^{1} (\tfrac{1}{i}) (U_{1-i,j} U^{y-1}_{0,j} yK^y_{PA}) \right] M_j k_p(T_j) \tag{30}$$

$$= \text{Rate}_{I,j} - (F_{R,j}/V_j) \, U_{1,j} + U_{00,1,j}/V_j$$

$$- \mathcal{L} U_{1,j} \, U_{0,j}^{y-1} \mathcal{J} \, M_j k_p'(T_j) + \mathcal{L} \sum_{i=0}^{1} \binom{1}{i}(U_{1-i,j} \, U_{0,j}^{y-1}) \mathcal{J}$$

$$M_j k_p'(T_j) \tag{31}$$

where $\binom{1}{i} = \dfrac{1!}{i!(1-i)!}$

Comparison of Experimental and Simulation Results

The simulation results depicted in Figs. 3 and 4 were obtained by integrating equations 12, 14 and 31 using the data in Table 1 to time one million seconds.

Table 1: Data for Simulation

Parameter	Value	Symbol
Initiator Feed Concentrations	$0.0914 \ \text{kgm}^{-3}$	I_o
Monomer Feed Concentrations	$2.782 \ \text{kgm}^{-3}$	M_o
Initiator Feed Flowrate	$0.667 \times 10^{-7} \text{m}^3/\text{s}$	$F_{I,j}$
Monomer Feed Flowrate	$0.583 \times 10^{-6} \ \text{m}^3/\text{s}$	$F_{M,j}$
Reactant Volume	$0.0027 \ \text{m}^3$	V_j
Reactor Temperature	$384 \ \text{K}$	T
Propagation Rate Constant	$.296 \ \text{m}^3 \, (\text{kg-mol})^{-1}\text{s}^{-1}$	k_p'
Initiation rate Constant	$1.95 \times 10^{-4} \ \text{m}^{3/2} (\text{kg-mol})^{-\frac{1}{2}} \text{s}^{-\frac{1}{2}}$	k_I'

The early experimental points in the concentration chain length distribution (Fig. 3) may be inaccurate.   They are calculated from the weight distribution obtained from the GPC.   The concentration chain length distribution is a function of the weight chain length distribution and the inverse of the chain length. Hence any error in the points in the weight chain length distribution is exaggerated.

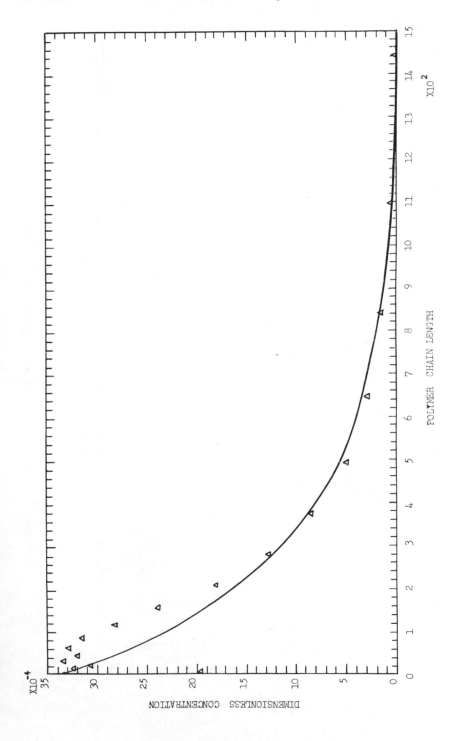

Figure 3. Differential concentration distribution ((△) experimental; (—) simulation)

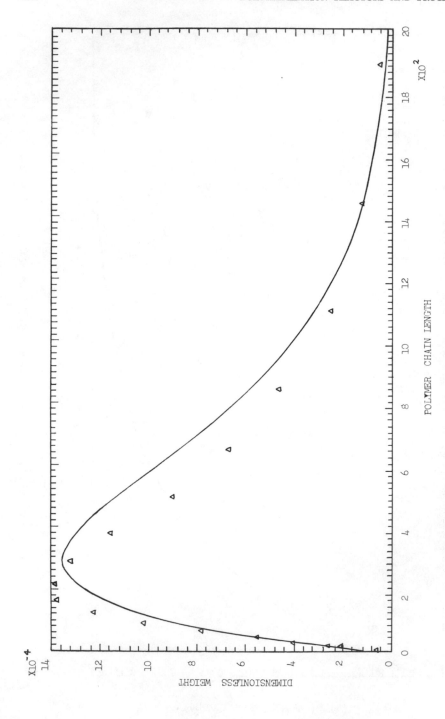

Figure 4. Differential weight distribution ((△) experimental; (—) simulation)

It can be seen that the theoretical and experimental curves agree well. The means also concur.

Conclusions

The described experimental rig for the anionic polymerisation of dienes has been shown to behave as an ideal CSTR. The mathematical model developed allows the prediction of the MWD at future points in the reactor history, once suitable kinetic parameters have been estimated.

Abstract

A pilot scale plant, incorporating a three litre continuous stirred tank reactor, was used for an investigation into the n-butyl lithium initiated, anionic polymerization of butadiene in n-hexane solvent. The rig was capable of being operated at elevated temperatures and pressures, comparable with industrial operating conditions.

Mathematical models of the reaction system have been developed, enabling prediction of the molecular weight distribution, based on the experimental data obtained from the pilot plant using on-line computer techniques. Results of simulation studies are compared with actual plant runs, and show a good measure of agreement.

Literature Cited

1. Gilman, H., Haubein, M. J. Am. Chem. Soc. (1944), 66, 1515.

2. Gear, G.W. "Numerical Initial Value Problems in ODEs". Prentice-Hall, New Jersey, 1971.

3. Dew, P.M., West, M.R. Department of Computer Studies, Leeds University, Report 107 (1978).

4. Dew, P.M., West, M.R. Department of Computer Studies, Leeds University, Report 111 (1978).

5. Elderton, W.P. "Frequency Curves and Correlation". Cambridge University Press, 1938.

6. Bamford, C,H., Tompa, H. Trans. Faraday Soc. (1954), 50, 1097.

RECEIVED February 19, 1979.

Anionic Styrene Polymerization in a Continuous Stirred-Tank Reactor

MICHAEL N. TREYBIG[1] and RAYFORD G. ANTHONY

Department of Chemical Engineering, Texas A&M University, College Station, TX 77843

In the design, optimization, or control of a polymerization reactor, a mathematical model which adequately represents the process is desirable. In the formulation of such a model, information is required on both the kinetics of the specific reaction and the mixing pattern of the reaction vessel used. For continuous stirred tank reactors, the assumption of perfect or micro-mixing is frequently made and the corresponding design equations used to estimate the reactor's performance. However, in many large scale industrial polymerization processes the occurrence of imperfect mixing or segregation is more probable. In the case of a segregated polymerization reactor, design equations are required which give a different molecular weight distribution from that obtained for the micro-mixed case. Since the processibility and mechanical properties of a polymer fraction are strongly dependent on the shape of the molecular weight distribution, it is important to know the effects of imperfect mixing on the shape of the molecular weight distribution and the degree of imperfect mixing occurring in a reactor.

Scope and Objectives

The objectives of this work were: to study the effect of segregated mixing in a stirred tank flow reactor on the molecular weight distribution of polystyrene; to determine the degree of segregation, if any, occurring in a bench scale laboratory reactor; and to evaluate the usefulness of reactor flow models based on micro- and macro-mixing in a constant-flow, stirred-tank reactor. Styrene was polymerized in a bench scale laboratory reactor with polystyryllithium seed in benzene solvent. A seeded polymerization system was chosen to simplify the kinetic description of the process compared with a system involving simultaneous initiation and propagation reactions. Mathematical models based on concepts of micro- and macro-mixing in a stirred tank reactor were de-

[1]Current address: Shell Development Company, Westhollow Research Center, Houston, Texas

0-8412-0506-x/79/47-104-295$08.00/0

veloped. These models utilize kinetic descriptions of this polymer system from previous studies of the system, as well as data
obtained in this investigation. Results from the laboratory
experimentation and mathematical simulation were compared. The
comparison was used to determine the suitability of the mathematical simulation for modeling the polymerization process.

Theory

Reaction Mechanism. The reaction mechanism of the anionic-
solution polymerization of styrene monomer using n-butyllithium
initiator has been the subject of considerable experimental and
theoretical investigation (1-8). The polymerization process
occurs as the alkyllithium attacks monomeric styrene to initiate
active species, which, in turn, grow by a stepwise propagation
reaction. This polymerization reaction is characterized by the
production of straight chain active polymer molecules ("living"
polymer) without termination, branching, or transfer reactions.
    The stoichiometry of the polymerization process may be rep-
resented by the simple reaction scheme:

$$I + M \rightarrow P_1 \tag{1}$$

$$P_1 + M \rightarrow P_2 \tag{2}$$

$$P_j + M \rightarrow P_{j+1} \quad j = 2, \infty \tag{3}$$

However, the mechanisms by which the initiation and propagation reactions occur are far more complex. Dimeric association
of polystyryllithium is reported by Morton, et al. (9) and it is
generally accepted that the reactions are first order with respect
to monomer concentration. Unfortunately, the existence of associated complexes of initiator and polystyryllithium as well as
possible cross association between the two species have negated
the determination of the exact polymerization mechanisms (8, 10,
11, 12, 13). It is this high degree of complexity which necessi-
tates the use of empirical rate equations. One such empirical
rate expression for the auto-catalytic initiation reaction for the
anionic polymerization of styrene in benzene solvent as reported
by Tanlak (14) is given by:

$$R_I = k_I I M (1 + \phi P_T^3) \tag{4}$$

Tanlak found the following relations for the propagation
reactions and monomer consumption:

$$R_{Pj} = \alpha(P_{j-1} - P_j)M \tag{5}$$

$$R_P = \alpha P_T M \tag{6}$$

$$R_M = R_I + R_P \tag{7}$$

where:

$$\alpha = \frac{kp}{\frac{1}{2} + \sqrt{\frac{1}{4} + 2K_P^{-1} P_T}} \tag{8}$$

Similar results for the propagation reactions were obtained by Timm and Kubicek (15).

In this work, the characteristic "living" polymer phenomenon was utilized by preparing a seed polymer in a batch reactor. The seed polymer and styrene were then fed to a constant flow stirred tank reactor. This procedure allowed use of the lumped parameter rate expression given by Equations (5) through (8) to describe the polymerization reaction, and eliminated complications involved in describing simultaneous initiation and propagation reactions.

Mixing Models. The assumption of perfect or micro-mixing is frequently made for continuous stirred tank reactors and the ensuing reactor model used for design and optimization studies. For well-agitated reactors with moderate reaction rates and for reaction media which are not too viscous, this model is often justified. Micro-mixed reactors are characterized by uniform concentrations throughout the reactor and an exponential residence time distribution function.

The concept of a well-stirred segregated reactor which also has an exponential residence time distribution function was introduced by Dankwerts (16, 17) and was elaborated upon by Zweitering (18). In a totally segregated, stirred tank reactor, the feed stream is envisioned to enter the reactor in the form of macromolecular capsules which do not exchange their contents with other capsules in the feed stream or in the reactor volume. The capsules act as batch reactors with reaction times equal to their residence time in the reactor. The reactor product is thus found by calculating the weighted sum of a series of batch reactor products with reaction times from zero to infinity. The weighting factor is determined by the residence time distribution function of the constant flow stirred tank reactor.

Many mixing models which utilize the simplified concepts of micro-mixing and segregation have been introduced. Most notable of these are the two-environment models of Chen and Fan (19), Kearns and Manning (20), and others (21, 22), and the dispersion models of Spielman and Levenspiel (23), and Kattan and Adler (24).

Since polymerization reactions in continuous stirred tank reactors are often carried out under conditions of high viscosity not conducive to micro-mixing, theoretical and experimental investigations have been made to determine the effects of segregation on the molecular weight distribution for various polymer systems (25, 26, 27, 28). Ahmad (27) studied the effect of mixing on the molecular weight distribution of polyisoprene and Tadmor and Biesenberger (28) studied the effect of segregation on

molecular weight distributions. For a stepwise addition non-
terminating polymerization in a segregated constant flow stirred
tank reactor, these authors found that a polymer would be produced with a molecular weight distribution that is broader than
that of a batch reactor but more narrow than that of a micro-mixed
reactor.

The design equations for the mixing conditions considered in
this paper are presented in Table I. The micro-mixed model was
modified to include the effect of inactive or dead polymer and
the effect of part of the reactants passing through the reaction
zone without reacting. The equations for the well-stirred segregated reactor and for the batch reactor are also presented in
Table I. Figure 1 illustrates the growth characteristics of
polymer chains in micro-mixed and segregated well-stirred reactors.
For the micro-mixed CFSTR the growth lines from a seed polymer
are linear, while from the segregated CFSTR they exhibit curvature
due to the change of monomer concentration as a segregated lump
passes through the reactor.

Experimental

Reactor Design. The continuous polymerization reactions in
this investigation were performed in a 50 ml pyrex glass reactor.
The mixing mechanism utilized two mixing impellers and a Chemco
magnet-drive mechanism.

The glass reactor, shown in Figure 2, has single inlet and
outlet ports and one thermocouple port. The reactor shell is made
from a section of pyrex tubing 4.4 cm OD and 4.0 cm ID. The inlet
and outlet ports are made from 1/4 in OD x 1.0 mm ID capillary
tube. The thermocouple port is made from 1/4 in OD x 5/32 in ID
glass tubing. Glass to stainless connections are made using 1/4
in stainless Swagelok fittings with Teflon front ferrules and a
1/4 in x 0.065 in viton '0'-ring. The stainless fittings used at
the inlet and thermocouple ports are 1/4 in to 1/16 in reducers.
The inlet port fitting is connected by a short section of 1/16 in
tubing to a 1/16 in stainless tee which is used to pre-mix the
monomer and living-polymer feed streams. The thermocouple port
allows entrance of a type "T" thermocouple in a 1/16 in stainless
sheath. To fill the void between the thermocouple and the tubing
wall of the thermocouple port, a plug of Teflon 5/32 in OD with
a 1/16 in ID axial hole is placed in the thermocouple port. The
fitting used at the reactor outlet port is a 1/4 in to 1/8 in
reducer and is connected to the reactor effluent line of 1/8 in
teflon tubing.

Two impellers are included in the reactor configuration shown
in Figure 2. A three-bladed turbine with 45° pitch and blades 1/8
in x 5/8 in is mounted on the impeller shaft at the top of the
reactor. A three-bladed propeller with 45° pitch is mounted at
the bottom of the impeller shaft at approximately two-thirds of
the reactor depth.

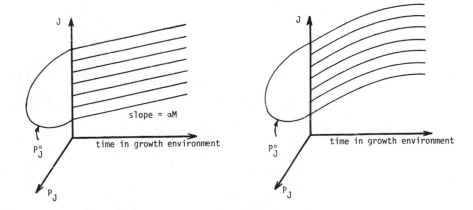

Figure 1. Growth characteristics for seed polymer in CFSTR environments: (a) growth characteristics for polymer chains in a micro-mixed environment; (b) growth characteristics for polymer chains in a segregated environment

TABLE I.   REACTOR MIXING MODELS

The derivations of the equations are given by Treybig (32)

Micro-mixed CSTR	Inactive (Dead) Polymer in Micro-mixed CFSTR
1. $M = \dfrac{M^o}{1 + \alpha\Theta P_T}$	1. $M = \dfrac{M^o}{1 + \alpha\Theta\phi_A P_T}$
2. $P_{j_{min}} = \dfrac{P_{j_{min}}}{1 + \alpha M\Theta}$	2. $P_{j_{min}} = \dfrac{P^o_{j_{min}}}{1 + \alpha M\Theta\phi_A}$
3. $P_j = P_{j_{min}} \exp\left[t_{jm}/\Theta\right] + \displaystyle\int_0^{\tau_{jm}} P_j^o(\tau)\,d\tau$	3. $P_j = P_{Dj}^o + P_{Aj}$
$P_j^o(\tau) = P_j^o(t)$	$P_{Aj} = P_{Aj_{min}} \exp[-t_{jm}/\Theta] + \dfrac{1}{\Theta}\displaystyle\int_0^{t_{jm}} P_{Aj}(t)\,\exp[-t/\Theta]\,dt$
$t = -\Theta\ln(1 - \tau)$	$P^o_{Aj}(t) = P^o_{(j+\alpha M\phi_A t)}$
$\tau_{jm} = 1 - \exp\left[-\left(\dfrac{J - j_{min}}{\alpha M}\right)\right]$	$t_{jm} = \dfrac{J - j_{min}}{\alpha M\phi_A}$
$t_{jm} = \dfrac{J - j_{min}}{\alpha M\Theta}$	
4. $D_n = D_n^o + \alpha M\Theta$	4. $D_n = D_n^o + \alpha M\Theta\phi_A$
5. $D_w = D_w^o \dfrac{D_n^o}{D_n} + \left(1 - \dfrac{D_n^o}{D_n}\right)\left(1 + 2D_n\right)$	5. $D_w = D_w^o \dfrac{D_n^o}{D_n} + \left(1 - \dfrac{D_n^o}{D_n}\right)\left[1 + 2\left(\dfrac{D_n - D_n^o}{\phi_A}\right) + 2D_n^o\right]$
6. $\dfrac{D_w}{D_n} = D_w^o \dfrac{D_n^o}{D_n^2} + \left(1 - \dfrac{D_n^o}{D_n}\right)\left(\dfrac{1}{D_n} + 2\right)$	6. $\dfrac{D_w}{D_n} = D_w^o \dfrac{D_n^o}{D_n^2} + \left(1 - \dfrac{D_n^o}{D_n}\right)\left[\dfrac{1}{D_n} + 2\left(\dfrac{D_n - D_n^o}{\phi_A D_n}\right) + 2\dfrac{D_n^o}{D_n}\right]$

TABLE I. (CONTINUED)

By-Passing in Micro-mixed CSTR

1. $M = \dfrac{M^o(1 + \alpha \Theta_R P_T \phi_B)}{1 + \alpha \Theta_R P_T}$

2. $P_j = \phi_B P_{Bj}^o + \phi_R P_{Rj}$; $j \geq j_{min}$

3. $P_{j_{min}} = \dfrac{P_{j_{min}}^o}{1 + \alpha M \Theta_R} \left[\exp\left[\dfrac{-t_{jm}}{\Theta_R}\right] + \int_0^{t_{jm}} P_{Rj}(t) \dfrac{1}{\Theta_R} \exp\left[\dfrac{-t}{\Theta_R}\right] dt \right]$

$P_{RJ} = P_{Rjmin}$

$J(t) = j + \alpha M_R t$

$t_{jm} = \dfrac{J - j_{min}}{\alpha M_R}$

Equations for D_n, D_w and D_w/D_n are same as for equations for dead polymer with ϕ_A replaced by ϕ_R.

Well-Stirred Segregated CSTR

1. $M_{seg} = M_{Micro} = \dfrac{M^o}{1 + \alpha P_T \Theta}$

2. $(P_j)_{seg} = \int_0^\infty P_j(t)_{Batch} \dfrac{dE(t)}{dt} dt$

3. $(P_j)_{seg} = \int_0^\infty P(j - \Delta j(t), 0) \dfrac{dE(t)}{dt} dt$

4. $\lambda_{i_{seg}} = \int_0^\infty \lambda_i(t)_{Batch} \dfrac{dE}{dt} dt$; $i = 0, 1, 2$

5. $D_n = \lambda_1/\lambda_0$, $D_w = \lambda_2/\lambda_1$

6. $\dfrac{D_w}{D_n} = \dfrac{\lambda_2 \lambda_0}{\lambda_1^2}$

TABLE I. (CONTINUED)

Batch Reactor

1. $M(t)_{Batch} = M^o exp[-\alpha P_T t]$

2. $P_{j_{min}}(t) = P^o_{j_{min}} exp[M^o/P_T(exp(-\alpha P_T t) - 1)]$

3. $P_j(t) = P^o_j (J - \Delta j(t), 0)$

$j(t) = j^o + \int_0^t \alpha M(t) dt$

$\Delta j(t) = \left\{ M^o/P_T \right\} [1 - exp(-\alpha P_T t)]$

4. $\lambda_0(t) = \lambda^o_0$

5. $\lambda_1(t) = \lambda^o_1 + \Delta j(t) \; \lambda^o_0$

6. $\lambda_2(t) = \lambda^o_2 + 2\Delta j(t)\lambda^o_1 + [\Delta j(t)]^2 \lambda^o_0$

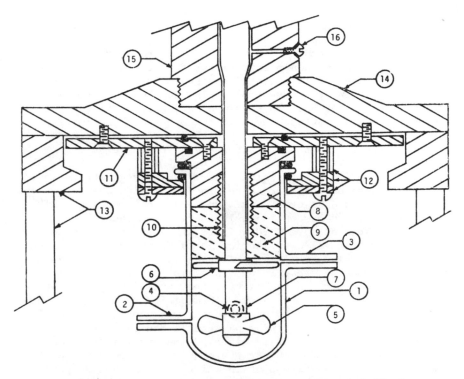

Figure 2.   Diagram of laboratory reactor: (1) glass reactor vessel; (2) inlet port;
(3) outlet port; (4) thermocouple port; (5) propeller impeller; (6) turbine impeller;
(7) impeller shaft; (8) stainless steel center plug; (9) Teflon center plug; (10) center
bolt; (11) interface plate; (12) flange assembly; (13) Chemco reactor support; (14)
Chemco reactor top closure plate; (15) Chemco reactor impeller shaft bearing
housing; (16) reactor blead port

In order to use the air-powered Chemco magnet drive mechanism, certain modifications to the existing unit are required along with additional interfacing assembly. The modifications and interface assembly are shown in Figure 2. To avoid shortening the original impeller shaft and to maintain a length to diameter ratio in the reactor close to unity, the glass reactor is made 9.5 cm in length and a stainless and Teflon plug extending 4.25 cm into the reactor is required. The plug is mounted on an interface plate which is mounted to the bottom of the standard Chemco assembly. To mount the interface plate to the Chemco reactor, four additional holes were drilled and tapped. To mount the glass reactor, a flange assembly with six bolts and spacers was constructed. Viton rubber '0'-rings are used to maintain liquid seals in the flange assembly and in the plug interface assembly. In addition, a port is drilled in the lower half to the impeller shaft bearing housing. This port is used to flush the void space between the impeller shaft and the surrounding assembly with liquid from the reactor. A seal is provided by a small screw and a Teflon washer.

Residence Time Distribution. Residence time distribution experiments were performed to determine the mixing speeds (rpm) at which well-mixed reactor conditions could be expected. RTD runs were performed by continuously feeding benzene as well as polymer solutions of 15 and 30 weight percent polystyrene in benzene. The exit age distribution for a given rpm and average residence time was determined by introducing an impulse change (styrene was used as the pulsing media) in the refractive index of the material entering the reactor and continuously monitoring the exit stream refractive index. In addition to measuring the exit age distribution, a color indicator was used to give a visual indication of the mixing pattern in the reactor.

To run the residence time distribution experiments under conditions which would simulate the conditions occurring during chemical reaction, solutions of 15 weight percent and 30 percent polystyrene in benzene as well as pure benzene were used as the fluid medium. The polystyrene used in the RTD experiment was prepared in a batch reactor and had a number average degree of polymerization of 320 and a polydispersity index, DI, of 1.17. A curve showing the differential weight fraction versus chain length for the batch-prepared polystyrene is given in Figure 3. Figure 4 illustrates the exit age distributions obtained from the RTD experiments with benzene and with a 30 weight percent polymer solution.

Preparation of Reactants. Reagent-grade, thiophene free benzene was stored over 4A molecular sieves and sodium ribbon in a helium atmosphere. Styrene was distilled to remove dissolved oxygen and moisture and stored under a helium atmosphere. Prior to use, styrene was injected into a copious quantity of methanol to determine if any polymerization had occurred. If there was

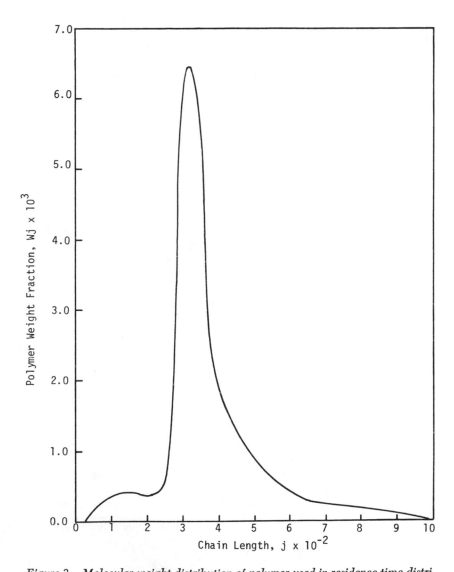

Figure 3.   Molecular weight distribution of polymer used in residence time distribution tests

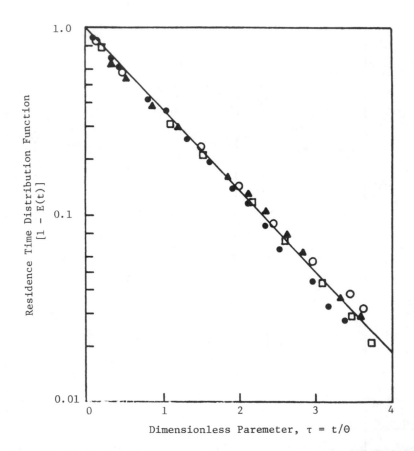

Figure 4. Residence time distribution for the 50-mL glass reactor with final mixing configuration (30 wt % Polystyrene: (θ) 10.0 min; (—) theoretical; (●) 1000 rpm; (▲) 750 rpm. Benzene: (θ) 10.0 min; (○) 1000 rpm; (□) 500 rpm)

no precipatant, the styrene was used. The initiator, n–butyllithium, which was used to prepare the seed was obtained from
Lithium Corporation of America as a 2.0 N butyllithium solution
in hexane. The concentration of butyllithium was determined using
a modification of the procedures presented by Uranek (29) and
Kolthoff (30).

Three polymer seeds were prepared in a batch reactor. The
reactor with styrene and benzene was cooled to $0^{\circ}C$ in an ice bath,
initiator was injected into the reactor and reaction began with a
gradual increase in temperature. Table II presents the initial
conditions used in preparing the seed polymer and the molecular
weights of the seed polymer. The molecular weight distribution
of the polymer seeds are shown in Figure 5.

Analysis of Data. Gel Permeation Chromatography (GPC) was
used to determine the molecular weights and molecular weight distribution of the seed and polymers that were produced. A modified
version of Smith's (31) method was used to convert the GPC data to
molecular weights and molecular weight distribution.

The concentration of polymer in the seed was determined by
modification of the method of Kolthoff (30) and Uranek (29) and
from the GPC data. The results agreed within 20% which was considered to be within experimental error. However, the GPC value
obtained as follows was used in all subsequent calculations

$$P_{T_{seed}} = \frac{(M_o)_{batch}}{(D_n)_{seed}} \tag{9}$$

The use of this equation to determine the polymer concentration
in the seed assumes that all of the styrene was reacted.

The styrene conversion for the continuous flow stirred tank
experiments was determined utilizing the concentration of the
polymer in the feed and the number average degrees of polymerization

$$X_m = \frac{P_T(D_n - D_n^o)}{M^o} \tag{10}$$

Styrene conversion calculated by this equation and styrene conversion obtained for runs 12–15 by gravimetric methods were in good
agreement. In general, the gravimetric technique was 1 to 5%
points greater than conversions calculated using the GPC data.

More detail on the experimental technique and procedures is
given by Treybig (32).

Experimental Apparatus. The experimental apparatus used in
the continuous polymerization reactions of this investigation was
constructed and used by Ahmad (27) for earlier studies of isoprene

TABLE II. INITIAL CONDITIONS FOR BATCH SEED PREPARATION
AND RESULTS OF GPC ANALYSIS OF AVERAGE MOLECULAR WEIGHTS

Seed No.	Initial Monomer Concentration (moles/kgm)	Initial BuLi Concentration (moles/kgm)	$\overline{M}_n$	$\overline{D}_n$	$\dfrac{\overline{M}_w}{\overline{M}_n}$	Used In Runs Nos.
I	1.508	0.0135	15,725	151	1.42	1
II	1.396	0.0128	15,309	147	1.36	2–11
III	1.354	0.0125	13,851	133	1.28	12–15

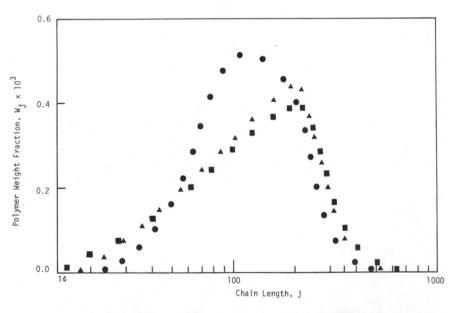

Figure 5. *Weight fraction distributions for polymer Seeds I (●), II (▲), and III (■)*

polymerization reactions. A modification to the original apparatus was to use the 50 ml glass reactor and mixing scheme described
in the Reactor Design section of this paper. A diagram of the
experimental apparatus given by Ahmad (27) is shown in Figure 6.

Helium over pressure of 40 to 50 psig in the feed vessels was
used to force the liquid feeds through rotameters and into the
reactor. Reactor feed lines of 1/8 in stainless steel were
equipped with 10 psi check valves, stainless needle valves and
Brooks Sho-rate rotometers (R-D-15D, seed; R2-15AAA, monomer) to
control and meter the feed rates. The temperature of the reactor
bath was maintained using a Sargent Model "T" thermometer equipped
with a 300 watt heating element. The temperature inside the reactor was monitored using a type "T" thermocouple connected to a
Newport Model 2400 digital voltmeter. For a more detailed description of the apparatus, see Ahmad (27).

Reactor Conditions for Experimental Runs. Operating conditions for the continuous, stirred tank reactor runs were chosen
to study the effects of mixing speed on the monomer conversion
and molecular weight distribution at different values for the
number average degree of polymerization of the product polymer.
It was believed that micro-mixing would be favored at conditions
resulting in small increases in the degrees of polymerization of
the product polymer over that of the seed polymer, whereas, larger
increases might be more favorable to segregated mixing. Before
running the laboratory experiments, simulations of polymerization
runs were performed to determine operating conditions which would
give the proper degree of polymerization as well as to illustrate
the effects of micro-mixing and segregation on the product distribution. The results of one simulation, using the kinetic
equations of Tanlak (14) are shown in Figure 7. The experimental
operating conditions for the continuous reactor runs of this investigation are given in Table III. Runs 1 through 11 were at
approximately 25°C. For runs 12 through 15 the reaction temperature was increased to approximately 35°C to increase the reaction
rate and thereby increase the likelihood of observing the effects
of segregation.

The utilization of the seed mixture was as follows: Run 1
used Seed I, Runs 2-11 used Seed II, and Runs 12-15 used Seed III.

Experimental Procedure. For the initial start-up of the
continuous tirred tank reactor, the mixing speed and bath temperature were adjusted with the reactor full of solvent. The polymer
seed and monomer feed rates were then adjusted simultaneously.
The feed flow rates and the reactor and bath temperatures were
monitored at five minute intervals. After five to six residence
times, two samples of the reactor effluent were collected in 50 ml
Erlenmeyer flasks containing approximately 20 ml of benzene saturated with water. Sufficient polymer solution was collected to
give a 3.0 weight percent polymer solution. The samples were

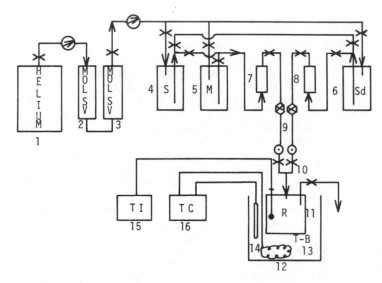

Figure 6. Experimental apparatus for continuous polymerization of styrene:
(⊙) Needle valve; (×) valve; (ⓐ) pressure regulator; (⊗) check valve; (—•) ther-
mocouple; (1) compressed helium; (2, 3) molecular sieve columns; (4, S) benzene
(solvent) tank; (5, M) styrene (monomer) tank; (6, Sd) seed polymer tank; (7, 8)
rotameters; (9) teflon tubing; (10) premixing tee; (11, R) reactor; (12) temperature
bath; (13) heating coil; (14) thermometer probe; (15, TI) temperature indicator
(16, TC) temperature controller

TABLE III.   OPERATING CONDITIONS FOR CFSTR RUNS 1–15

Run No.	Feed Concentration of Styrene, M° (gm-mole/liter)	Feed Concentration of Seed Polymer, P_T (gm-mole/liter)	Mean Residence Time (min)	Mixing Speed (RPM)	Reaction Temperature (°C)
1	0.856	0.0081	25.4	1000	24.8±0.25
2	0.685	0.0075	31.6	1000	23±2
3	0.729	0.0075	31.6	500	22±2
4	3.583	0.0034	22.4	1000	26.7±0.25
5	2.309	0.0052	19.5	1000	22±2
6	2.309	0.0052	19.9	500	20±2
7	2.308	0.0052	19.6	750	20±2
8	3.456	0.0035	15.3	1000	25.7±0.25
9	3.458	0.0035	15.4	500	26.2±0.25
10	4.147	0.0024	15.3	1000	25.6±0.25
11	4.157	0.0024	14.6	500	26.4±0.25
12	2.678	0.0067	21.9	1000	36±1
13	2.714	0.0066	22.2	500	37±1
14	2.656	0.0066	21.7	300	38±1
15	2.675	0.0066	21.3	250	35±0.25

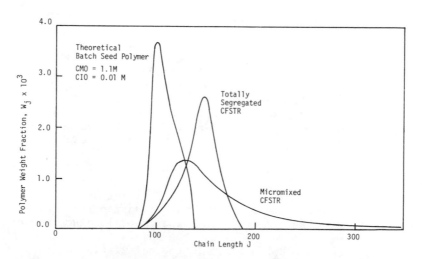

Figure 7.   Theoretical polymer distributions, based on kinetic description of Tan-lak (14) for micro-mixed and totally segregated CFSTRS with polymer feed (CFSTRS: CMO = 0.5M; PT = 0.01M; θ = 20.0 min; XM = 0.70)

later filtered and then analyzed using the GPC. After the polymer
samples were taken, the impeller was stopped to allow estimation
of the volume of gas which collected in the reactor (due to degassing of helium from the feed stream) during the run. This
reduction in the effective reaction volume of the reactor was
noted and the gas was removed from the reactor through the exit
port by tilting the reactor. Subsequent runs were then made by
adjusting the feed flowrates and then the mixing speed with the
reactor initially filled the reaction medium from the previous
run.

Results

The experimental monomer conversion and degrees of polymerization for the continuous reactor runs are given in Table IV.

Experimental Values for the Lumped-Parameter Propagation
Rate Constant. The experimental values for the lumped-parameter
propagation rate constant were determined assuming a micro-mixed
reactor, styrene concentration and solving for α. The results for
Runs 1-15 are included in Table V. The value for the propagation
constant based on a segregated model are the same as that for a
micro-mixed model. For the case of a micro-mixed reactor with
dead-polymer or a micro-mixed by-pass reactor, the true value of
α would be larger than the value reported for the micro-mixed case
by factors of $1/\phi_D$ and $1/[1 - \alpha\Theta P_T \phi_B/\Theta_R]$, respectively. This
would compensate for the decrease in monomer conversion associated
with dead-polymer and by-passing.

Calculated Degrees of Polymerization. The calculated degrees
of polymerization for the micro-mixed, segregated, and micro-mixed
reactor with dead-polymer models are given in Table VI. Values
for the lumped parameter propagation rate constant used in the
simulations were calculated such that the monomer conversions for
the models would be the same as that for the laboratory reactor.
Therefore, the number average degrees of polymerization for each
model is equal to the experimentally observed number average. For
the micro-mixed reactor with dead-polymer model, average values
of the fraction dead-polymer, $\phi_{D_{Avg}}$, were used for each of the
different seed mixtures. (Note that $\phi_D = 1 - \phi_A$.)

The average values of ϕ_D for each seed were determined by
averaging the values of ϕ_D required to match the experimental
number and weight average degrees of polymerization. The value of
ϕ_D for each run was found by solving the equation for D_w/D_n in
Table I for ϕ_D and substituting the experimental values for the
average degrees of polymerization. The values ϕ_D calculated for
each run are given in Table VII. In the calculation of $\phi_{D_{Avg}}$ for
Seed II the values of ϕ_D for Runs 3 and 4 were not used. The
values for $\phi_{D_{Avg}}$ are given in Table VIII.

TABLE IV.   EXPERIMENTAL MONOMER CONVERSIONS AND
DEGREES OF POLYMERIZATION FOR CFSTR RUNS 1-15

Run Number	rpm	X_m	D_n	D_w	D_w/D_n
1	1000	0.524	206	289	1.41
2	1000	0.586	200	267	1.34
3	500	0.411	187	263	1.41
4	1000	0.585	768	1411	1.84
5	1000	0.446	346	579	1.67
6	500	0.446	346	547	1.58
7	750	0.428	338	558	1.65
8	1000	0.377	524	995	1.90
9	500	0.423	568	1061	1.87
10	1000	0.311	586	1598	2.33
11	500	0.333	726	1701	2.34
12	1000	0.726	424	691	1.63
13	500	0.704	421	698	1.66
14	300	0.698	412	677	1.64
15	250	0.630	387	609	1.57

TABLE V.   EXPERIMENTAL VALUES FOR LUMPED PARAMETER PROPAGATION
RATE CONSTANT BASED ON MICRO-MIXED CFSTR MODEL

Run Number	Rate Constant, α	Run Number	Rate Constant* α
1	5.31	9	13.74
2	5.98	10	12.33
3	2.95	11	14.28
4	18.65	12	18.08
5	7.97	13	16.18
6	7.82	14	15.91
7	7.40	15	12.09
8	11.46		

* α is a function of temperature and polymer concentration as given by Equation 3.

TABLE VI. COMPARISON OF EXPERIMENTAL AND CALCULATED DEGREES OF POLYMERIZATION

Run Number	Experimental			Micro-mixed CFSTR		Segregated CFSTR		Micro-mixed CFSTR With Dead Polymer Fraction	
	D_n	D_w	D_w/D_n	D_w	D_w/D_n	D_w	D_w/D_n	D_w	D_w/D_n
1	206	289	1.41	267	1.30	256	1.24	289	1.40
2	200	267	1.34	253	1.27	242	1.21	264	1.32
3	187	263	1.41	237	1.27	233	1.24	244	1.31
4	768	1411	1.84	1281	1.67	910	1.18	1687	2.20
5	346	579	1.67	484	1.40	412	1.19	576	1.67
6	346	547	1.58	484	1.40	412	1.19	576	1.67
7	338	558	1.65	470	1.39	404	1.20	556	1.65
8	524	995	1.90	811	1.55	661	1.26	1030	1.97
9	568	1061	1.87	895	1.58	708	1.25	1147	2.02
10	686	1598	2.33	1122	1.64	919	1.34	1464	2.13
11	726	1701	2.34	1199	1.65	967	1.33	1573	2.17
12	424	691	1.63	636	1.50	467	1.10	696	1.64
13	421	698	1.66	630	1.50	468	1.11	690	1.64
14	412	677	1.64	614	1.49	458	1.11	671	1.63
15	387	609	1.57	567	1.47	438	1.13	617	1.60

(D_n) calculated $= (D_n)$ experimental   for all runs.

TABLE VII.   FRACTION DEAD POLYMER REQUIRED TO MATCH EXPERIMENTAL
DEGREES OF POLYMERIZATION USING A MICRO-MIXED
REACTION WITH DEAD POLYMER

Run Number	Fraction Dead Polymer, ϕ_D	Run Number	Fraction Dead Polymer, ϕ_D
1	0.427	9	0.211
2	0.328	10	0.360
3	0.599	11	0.352
4	0.115	12	0.121
5	0.292	13	0.146
6	0.217	14	0.144
7	0.291	15	0.112
8	0.253		

TABLE VIII.   AVERAGE FRACTION DEAD POLYMER FOR SEED MIXTURES

Seed Number	CFSTR Run Number	Average Fraction Dead Polymer, $\phi_{D_{Avg}}$
I	1	0.427
II	2 - 11	0.288
III	12 - 15	0.131

Calculated Molecular Weight Distributions. The calculated
weight fraction distributions for the micro-mixed, segregated,
and micro-mixed reactor with dead-polymer models for Runs 2, 5,
8, 10 and 12 are shown along with the experimental distributions
in Figures 8 through 12. These figures illustrate the effects of
micro-mixing and segregation on the weight fraction distribution
as well as the ability of the models to simulate the experimental
distributions at different degrees of polymeriztion. The calculated mole fraction distributions for Runs 8 and 12 are shown
with the experimental distributions in Figure 13 and 14.

Streaking Observed in Reaction Medium During Continuous
Polymerizations. Non-uniformities in the reactor contents in the
form of streaks were observed during continuous polymerizations
at mixing speeds of less than 1,000 rpm. At mixing speeds of
1,000 rpm, the reactor appeared to be divided into two homogeneous
mixing zones: one occupying the upper half of the reactor and
the other occupying the lower half. At lower mixing speeds for
Runs 3 and 7, a feed stream or streak was observed to pass from
the reactor feed port over the blades of the lower impeller and
down into the center of the lower impeller. Durings Runs 6, 8
and 13, additional streaks were observed in the lower mixing
zone. During Runs 9, 11, 14 and 15, considerable streaking was
observed in both the upper and lower mixing zones. It should
also be pointed out that during Run 11 (at low rpm and high degrees
of polymerization) in which an appreciable amount of degassing
occurred, small bubbles were occassionally observed to travel
from the top of the reactor into the lower mixing region. This
is an indication that a well-mixed condition was achieved to at
least a macroscopic level.

Discussion

The 50 ml glass reactor proved to be well-suited for the
procedures implemented in this investigation. The small size
of the reactor allowed efficient use of the materials required
for both the residence time distribution studies and for the continuous polymerization experiments. Visual inspection of the
reactor contents during operation proved valuable in determining
imperfections in the mixing pattern during RTD studies as well
as for observing streaking and bubble formation in the reactor
during polymerizations. The only disadvantage associated with the
small glass reactor is the increased care in handling required
over that of a small stainless steel reactor.

Errors in the Temperature Measurement During Polymerizations
Runs. The internal reactor temperatures measured during Runs 2,
3, 5, 6 and 7 were found to be inconsistent when a heat balance
was made on the reactor. Errors in these temperature measurements
may be due to increased resistance of the reactor thermocouple

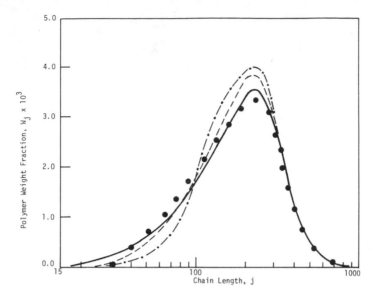

Figure 8. Comparison of experimental and calculated weight fraction distributions for Run 2 ((●) Exp; (——) Micro-ΦD; (— —) Micro; (·—·) Seg)

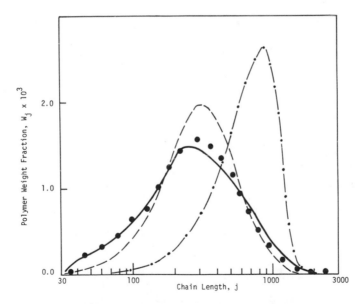

Figure 9. Comparison of experimental and calculated weight fraction distributions for Run 5 ((●) Exp; (——) Micro-ΦD; (— —) Micro; (·—·) Seg)

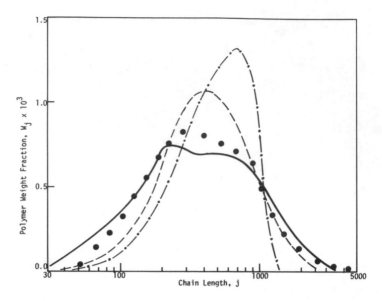

Figure 10.   Comparison of experimental and calculated weight fraction distributions for Run 8 (($\bullet$) Exp; (——) Micro-$^\Phi$D; (— —) Micro; ($\cdot$—$\cdot$) Seg)

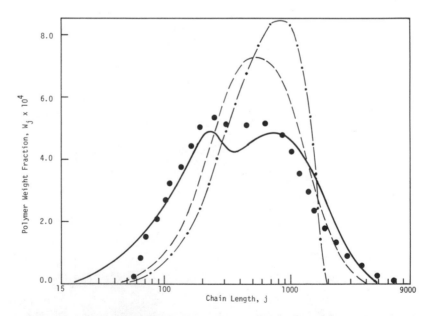

Figure 11.   Comparison of experimental and calculated weight fraction distributions for Run 10 (($\bullet$) Exp; (——) Micro-$^\Phi$D; (— —) Micro; ($\cdot$—$\cdot$) Seg)

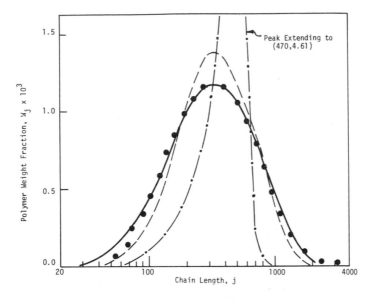

Figure 12.   Comparison of experimental and calculated weight fraction distributions for Run 12 ((●) Exp; (——) Micro-$^\Phi$D; (— —) Micro; (· — ·) Seg)

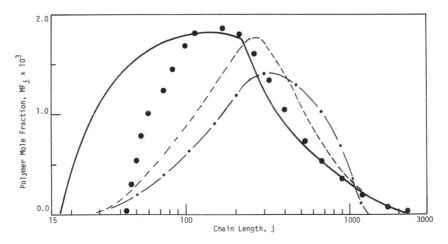

Figure 13.   Comparison of experimental and calculated mole fraction distributions for Run 8 ((●) exp; (——) Micro-$^\Phi$D; (— —) Micro; (· — ·) Seg)

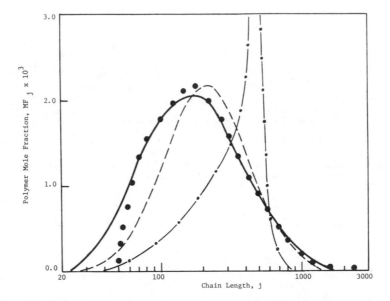

Figure 14. Comparison of experimental and calculated mole fraction distributions for Run 12 ((●) Exp; (——) Micro-$^\Phi$D; (— —) Micro; (·—·) Seg)

caused be bending the thermocouple sheath when handling the reactor between runs. Because of these errors the reaction temperatures for these runs are not known better than $\pm 2^{\circ}C$. Reactor Runs
12, 13 and 14 were intended to be carried out at $35^{\circ}C$, but it was
found that the temperature in the bath for the reference thermocouple had risen from $0.6^{\circ}C$ to $4.5^{\circ}C$ during the eight hour reaction
period required to complete the runs. The reaction temperatures
for Runs 12, 13 and 14 were corrected for this oversight and are
believed to be accurate to $\pm 1.0^{\circ}C$. The temperatures recorded for
the remainder of the reaction runs are accurate to $\pm 0.25^{\circ}C$.

Effect of Mixing Speed on Monomer Conversion and Molecular
Weight Distribution. The monomer conversion obtained in the
seeded polymerization of styrene in a CFSTR will be independent
of the degree of segregation and thus the mixing speed as long as
an exponential residence time distribution is maintained. Therefore, a dependence of monomer conversion on mixing speed for runs
with the same feed concentrations, average residence time, and
reaction temperature would indicate non-ideal mixing at the lower
mixing speeds. The experimental monomer conversions obtained for
runs at similar feed conditions and average residence times but
different mixing speeds are given in Table IV.

Initial comparison of CFSTR runs with similar feed conditions
indicates conditions for which the monomer conversion may be
dependent on mixing speed. However, when the effects of experimental error in monomer conversion and differences in reaction
temperature are considered, the monomer conversion is seen to be
relatively independent of mixing speed for rpm equal to or greater
than 500. Comparing Run 14 with Run 12 reveals a small decrease
in monomer conversion in spite of a rise in reactor temperature
of $2^{\circ}C$. This indicated the presence of a small amount of by-
passing or dead volume at the lower mixing speed. This imperfect
mixing pattern would also be present in Run 15.

The experimental molecular weight distributions given in
Figures 8 through 14 illustrate little or no significant effects
on the shape of the molecular weight distributions directly attributable to the mixing speed. Thus no effects of increased segregation with decrease in mixing speed were observed on the molecular
weight distributions.

Evaluation of Mixing Models. The micro-mixed reactor will
produce polymer distributions with increasing amounts of high
molecular weight tail as the degree of polymerization of the polymer product increases over that of the original seed polymer.
This trend is illustrated by the curves for the micro-mixed reactor
in Figures 8 through 14. Also characteristic of the seeded, micro-
mixed reactor is the convergence of the polydispersity index to 2
for a high degree of polymerization. This trend is illustrated to
some extent in Table VI which presents the calculated degrees of
polymerizations.

The micro-mixed reactor model was not able to simulate adequately the experimentally observed weight average degrees of
polymerizations or molecular weight distribution. These facts
are illustrated in Table VI and Figures 8 through 14. In general,
the weight average degrees of polymerization calculated for the
micromixed reactor were smaller than those observed experimentally.
This is due to the more narrow polymer distribution predicted by
the micro-mixed model as shown in Figures 8 through 14.

The micro-mixed reactor with dead-polymer model was developed
to account for the large values of the polydispersity index observed experimentally. The effect of increasing the fraction of
dead-polymer in the reactor feed while maintaining the same monomer conversion is to broaden the product polymer distribution and
therefore to increase the polydispersity index. As illustrated in
Table V, this model, with its adjustable parameter, ϕ_D, can
exactly match experiment average molecular weights and easily
account for values of the polydispersity index significantly
greater than 2.

The fair degree of consistency observed in the values of ϕ_D
for Seeds II and III and the excellent agreement between the
experimental molecular weight distribution and those calculated
with $\phi_{D_{Avg}}$, lends credibility to the dead-polymer model. The
agreement between experimental and calculated distribution at
increasing degrees of polymerization are given in Figures 8 through
14. The bimodal weight fraction distributions calculated for Runs
8 and 10, which are shown in Figures 10 and 11 are of particular
interest. There is good agreement between experiment and theory
in spite of limitations in the ability of the GPC data reduction
routine to handle bimodal distributions.

To differentiate between the micro-mixed reactor with dead-
polymer and the by-pass reactor models in this investigation, the
effect of mixing speed on the value of "ϕ" was observed. As
illustrated in Table V, the value "ϕ" is not observed to increase
with decreasing mixing speed as would be expected for a by-pass
reactor. This rules out the possibility of a by-pass model and
further substantiates the dead-polymer model.

The well-stirred segregated reactor will produce polymer
distributions with low molecular weight tails and sharp trunca-
tions at the high molecular weight ends at increased degrees of
polymerization of the polymer product. This is illustrated in
Figures 8 through 14. The value of the polydispersity index for
the segregated reactor product will always be less than that of
the micro-mixed reactor (assuming no dead-polmer) as illustrated
in Table VI.

The segregated model was not able to simulate the experimentally observed degrees of polymerization on the molecular
weight distributions. As shown in Figures 8 through 14, the
segregated distributions were in general too narrow and exhibited
peaks in the mole fraction and weight fraction curves which far
exceeded those observed experimentally.

Signficance of Streaking in Reaction Medium. Streaks in the reaction medium observed during most continuous polymerization runs indicate the presence of some degree of segregation or incomplete micro-mixing. But, as indicated in Figures 8 through 14 for the comparison of experimental and calculated distributions, no significant influence of segregation on the shape of the distributions was observed. In fact, the product distribution is simulated well using a micro-mixed model with dead-polymer. This anomaly may be explained in part by arguments due to Patterson (33) based on studies of a CFSTR using Monte Carlo techniques. The effects of micro-mixing on the molecular weight distribution are much more pronounced than those of segregation. According to Patterson (33) only a small increase in micro-mixing over that of total segregation will yield a polymer distribution very similar to that of micro-mixed reactor.

Conclusions

The most significant results and conclusions are summarized below:

1. The monomer conversion in this seeded polymerization system is independent of the degree of segregation as long as an exponential residence time distribution function is maintained.

2. The mixing speed had little or no signficant effect on the monomer conversions or the shape of the molecular weight distributions for mixing speeds of 500 rpm or greater.

3. A micro-mixed, seeded reactor will produce a broad polymer distribution with a high molecular weight tail and polydispersity index that approaches 2 at large degrees of polymerization.

4. The effect of dead-polymer and by-passing on the micro-mixed reactor for the same degree of monomer conversion is to broaden the product polymer distribution and thus allow values of the polydispersity index much larger than 2.

5. A well-stirred segregated reactor would produce a product polymer with a low molecular weight tail and a sharp truncation at the high molecular weight end for large degrees of product polymer polymerization. At equal monomer conversions the weight average degrees of polymerization will be less for a totally segregated reactor than for a micro-mixed reactor.

6. The micro-mixed reactor with dead-polymer model simulated the product of the laboratory reactor well within experimental accuracy.

7. In spite of visual indications of at least partial segregation, the concept of micro-mixing proved to be most useful in modeling the laboratory reactor.

Symbols

CFSTR	Constant flow stirred tank reactor
D_n	Number average degree of polymerization
D_w	Weight average degree of polymerization
$E(t)$	Exit age distribution, $dE/dt = \Theta^{-1}\exp(-t/\Theta)dt$
GPC	Gel Permeation Chromatograph
I	Initiator concentration
j, J	Polymer chain length
j_{max}	Largest polymer chain length in polymer distribution
j_{min}	Smallest polymer chain length in seed distribution and reactor effluent
k_I	Initiation rate constant
k_p	Propagation rate constant
K_p	Equilibrium constant for polymer association
M	Monomer concentration
MF_j	Mole fraction polymer of length j
$\bar{M}_n$	Number average molecular weight
$\bar{M}_w$	Weight average molecular weight
MWD	Molecular weight distribution
$P_j{}^o$	Concentration of polymer of chain length j in reactor feed
P_j	Concentration of polymer of chain length j
P_J	Concentration of polymer of chain length J
P_{AJ}	Concentration of active polymer of chain length j
P_{Bj}	Concentration of polymer of chain length j in the by-pass stream of a by-pass CFSTR
P_{Dj}	Concentration of dead-polymer of chain length j
P_{Rj}	Concentration of polymer of chain length j in the reactor zone of a by-pass CFSTR
P_T	Concentration of total polymer
R_I	Rate of initiation
R_M	Rate of monomer consumption
R_p	Rate of propagation
RTD	Residence time distribution
t	Time
t_{jm}	Time required for smallest polymer molecule to grow to length j
W_j	Weight fraction of polymer of length j
X_m	Monomer conversion
α	Lumped parameter propagation function
Δ	Denoting difference
Θ	Average residence time
Θ_R	Average residence time in reaction zone of a by-pass CFSTR
λ_0	Zeroth moment of polymer distribution
λ_1	First moment of polymer distribution
λ_2	Second moment of polymer distribution
τ	Dimensionless time defined as $[1 - \exp(t/\Theta)]$
ϕ	Autocatalytic rate constant for initiation
ϕ_A	Fraction active polymer in CFSTR with dead polymer
ϕ_B	Fraction by-pass in by-pass CFSTR

ϕ_D Fraction dead polymer in CFSTR with dead-polymer
ϕ_R Fraction passing through reaction zone in by-pass reactor
o   Superscript denotes feed stream of seed polymer

Acknowledgments

The authors appreciate the encouragement and support of this work by the Department of Chemical Engineering, the Texas Engineering Experiment Station, and E. I. duPont deNemours & Company.

Literature Cited

1. Worsfold, D. J. and Bywater, S., Makromol. Chem., 65, 245 (1963).
2. Bywater, S. and Worsofld, D. J., Can. J. Chem., 40, 1564 (1962).
3. Fetters, L. J., Ph. D. Dissertation, University of Akron, Ohio (1962).
4. Hsieh, H. L., J. Polymer Sci., A3, 153 (1965).
5. Hooke, R. and Jeeves, T. A., Assoc. Comp. Cach., 8, 2, 212 229 (1961).
6. O'Driscoll, K. F. and Tobolsky, A. V., J. Polymer Sci., 35, 259 (1959).
7. Johnson, A. F. and Worsfold, D. J., J. Polymer Sci., A3, 449 (1965).
8. Bywater, S. and Worsfold, D. J., Advan. in Chem. Ser., 52, 36 (1969).
9. Morton, M. and Fetters, L. S., J. Polymer Sci., A2, 3311 (1964).
10. Worsfold, D. J. and Bywater, S., Can. J. Chem., 38, 1891 (1960).
11. Margerison, D. and Newport, J. P., Trans. Faraday Soc., 59, 1891 (1963).
12. Hsieh, H. L. and Glaze, W. H., Rubber Chem. Technol., 43, 22 (1970).
13. Lenz, R. W., Organic Chemistry of Synethic High Polymers Interscience, New York (1967).
14. Tanlak, T., M. S. Thesis, Texas A&M University, College Station, Texas (1975).
15. Timm, D. C. and Kubicek, L. F., Chem. Eng. Sci., 29, 2145 (1974).
16. Danckwerts, P. V., Chem. Eng. Sci., 2, 1 (1953).
17. Danckwerts, P. V., Chemical Reaction Engineering, 12th Meeting, Eur. Fed. Chem. Eng., Amsterdam (1957).
18. Zwietering, T. N., Chem. Eng. Sci., 11, 1 (1959).
19. Chen, M. S. K. and Fan, L. T., Can. J. Chem. Eng., 49, 704 (1971).
20. Keairns, D. L. and Manning, F. S., AIChE J., 15, 660 (1969).
21. Ng, D. Y. C. and Rippin, D. W. T., Third Eur. Symp. Chem. Reaction Eng., Amsterdam, Pergamon Press, Oxford 161 (1965).

22. Goto, S. and Matsubara, M., Chem. Eng. Sci., 30, 61 (1975).
23. Spielman, L. A. and Levenspiel, O., Chem. Eng. Sci., 20, 247 (1965).
24. Kattan, A. and Adler, R. J., Chem. Eng. Sci., 27, 1013 (1972)
25. Nagasubramanian, K. and Graessley, W. W., Chem. Eng. Sci., 25, 1559 (1970).
26. Rao, D. P. and Edwards, L. L., Chem. Eng. Sci., 28, 1179 (1973).
27. Ahmad, A., Ph. D. Dissertation, Texas A&M University, College Station, Texas (1975).
28. Tadmor, Z. and Biesenberger, J. A., I&EC Fundamentals, 5, 336 (1966).
29. Uraneck, C. A., Burleigh, J. E. and Cleary, J. W., Anal. Chem., 40, 327 (1968).
30. Kolthoff, J. M. and Harris, W. E., Ind. Eng. Chem. Anal. Ed., 18, 161 (1946).
31. Lewin, S. Z., Chem. Educ., 43, Reprint (1966).
32. Treybig, M. N., "Effect of Mixing on Polymerization of Styrene," Master of Science Thesis, Texas A&M University, College Station, Texas (1977).
33. Patterson, G. K., Personal Communication (1977).

RECEIVED January 15, 1979.

Designing for Safe Reactor Vent Systems

LOUIS J. JACOBS, JR. and FRANCIS X. KRUPA

Monsanto Co., Corporate Engineering Dept., St. Louis, MO 63166

This paper is primarily concerned with safe venting of polymerization reactors, though the same principles apply to almost any vessel containing volatile, potentially hazardous substances. In polymerization vessels one usually deals with exothermic reactions of volatile monomers. The reactions may occur in either emulsion, suspension, mass or solution-type polymerization on a batch or continuous basis. Other papers at this conference have discussed each of these extensively and each has advantages and disadvantages regarding control of emergencies. The suspension and emulsion systems generally have a built-in heat sink with the water present, but exhibit higher vapor pressure due to the nearly additive effect of the immiscible monomer and water phases.

I. Defining the Venting Problem

The need for venting, or the cause of an emergency which results in a runaway reaction, can occur in several ways:

Cooling system failure could occur due to failure of pumps or controls supplying cooling media to the reactor vessel jacket, coils, or overhead reflux condensers. Piping to or from the condensers could become plugged or any of the heat exchange surfaces could become excessively fouled.

Agitator failure either due to electrical or mechanical failure could result in loss of system control and "hot spots" in the reactor. In suspension systems loss of agitation could negate much of the "heat sink" effect as the immiscible phases separate and stratify.

0-8412-0506-x/79/47-104-327$05.00/0

Incorrect vessel charge either due to automatic control failure or plant operator error could result
in excess catalyst or reactant concentration, etc.
This could cause a rapidly accelerating reaction
rate or could initiate unexpected side reactions,
which could be more severe than the normal reaction.

External fire could cause an emergency by overload-
ing the normal reactor systems that are operating
properly.

Each of these cases involves an accumulation of heat in
the system which manifests higher temperature and pressure. The increased temperature accelerates the reactions further which subsequently adds even more heat
to the system.

II. Strategies to Handle Emergencies

In the event that one or more of the cases cited
above will occur at some point in the life of a process, we need to have a design strategy to cope with
such emergencies. Selection of a strategy will involve
judgment of risks and likelihood of occurrence, which
will not be discussed here. There are several design
strategies that can be used to minimize the consequences of the emergency by anticipating system response.
Elaborate, redundant reactor control systems
could be installed, such as multiple temperature sens-
ing points. On high temperature, these trigger actions
such as feed shutdown, emergency cooling, or the addition of substances to deactivate the catalyst. Other
control techniques could include a high pressure switch
to activate automatically controlled venting by allowing volatiles to be vented from the reactor.
The quantity of volatile monomer present could be
limited by using smaller volume continuous reactors,
or using a semicontinuous monomer feed. Small quantities of monomers present would quickly be consumed by
an uncontrolled reaction, and with the system deprived
of further reactants pressure rise would be limited.
Another strategy would involve design of the
reactor vessel for a pressure rating in excess of any
likely emergency system pressure. This assumes we can
adequately predict all possible worst case situations,
which is doubtful.
A more conventional approach is to provide a safety relief valve or rupture disc to protect the vessel
by venting material when pressure approaches certain
limits, such as the maximum allowable working pressure.

This strategy may be used in combination with the first two strategies.

An alternate approach to the above is to provide parallel relief valve-rupture disc systems. The valve will have a setting slightly above the normal operating pressure with the rupture disc at about a 10% higher setting. The relief valve should control minor pressure excursions, can vent material and then reseat to minimize process losses. The rupture disc would provide the ultimate safety protection.

The remainder of this paper will discuss design of systems where venting of material is necessary.

III. Sizing the Vent System

A. Available Design Methods for Vent Sizing. Several methods are available to size the vent with a wide range of sophistication. The FIA chart, Fig. 1 prepared by the Factory Insurance Association in the mid 1960's is a simple chart summarizing a wealth of experience. Reactions are classed by the degree of exothermic reaction. With vessel size and a judgment of reaction type a vent size range can be selected. This chart was prepared to be a guide to insurance inspectors and not a design technique. Experience indicates, however, it is often used by designers to estimate a reactor vent size.

In 1967 a paper by Boyle (1) provided a more quantitative method for designing vents for polymer reactors. It was based on reaction rate, heat of reaction, and vapor pressure data. Boyle assumed that the venting of a system can be approximated by sizing to discharge the entire batch contents as a liquid. The vent line size is determined so the time to vent the entire batch contents is less than the time to go from relief set point to maximum allowable vessel pressure.

A frequent sizing technique, which is useful when the reaction kinetics and heat of reaction are not known, is to conduct small scale tests. Then scale up to large equipment is done by providing a vent with similar vent area per mass of contents.

In 1972 a paper on venting by Huff (2) documented concerns that many designers suspected: that to truly be safe the vent sizing of many systems should be based on assuming two-phase flashing flow in the vent system. A two-phase flow vent method developed by Huff was compared with Boyle's all-liquid method, and values from the FIA chart in Figure 2. It can be seen that under many conditions, previous methods were not

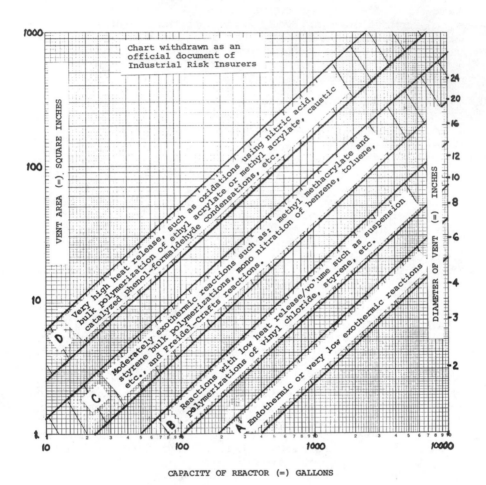

Figure 1.   FIA chart

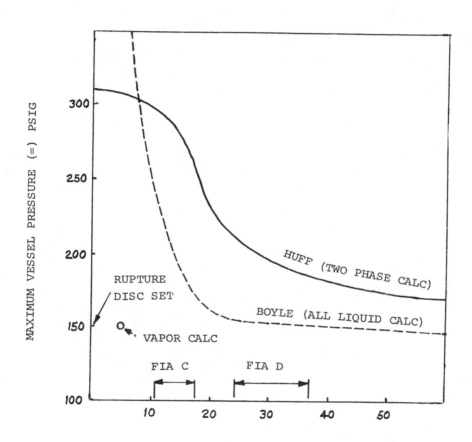

Figure 2.   Peak reactor pressure vs. relief line size (2)

providing conservative design.

Monsanto and other companies are working independently on design methods to size vents more rigorously
using two-phase flow calculations in complex computer
programs. Several assumptions have been made in an
effort to allow a wide range of application. Most
notable is the use of the correlations of Martinelli
and co-workers for pressure drop(3) and hold-up(4).
The momentum and energy balances are developed for the
separated flow regime by Hewlett and Hall-Taylor(5).
A homogeneous flow basis must be used when thermodynamic equilibrium is assumed. For further simplification it is assumed there will be no reaction occurring
in the pipeline. The vapor and liquid contents of the
reactor are assumed to be a homogeneous mass as they
enter the vent line. The model assumes adiabatic conditions in the vent line and maintains constant stagnation enthalpy for the energy balance.

The Martinelli correlations for void fraction and
pressure drop are used because of their simplicity and
wide range of applicability. France and Stein(6) discuss the method by which the Martinelli gradient for
two-phase flow can be incorporated into a choked flow
model. Because the Martinelli equation balances fric-
tional shear stresses and pressure drop, it is important to provide a good viscosity model, especially for
high viscosity and non-Newtonian fluids.

As the gas-liquid mixture travels down the vent
line, the phases will slip past each other and the
fluids will accelerate. This contribution to the
energy balance can be most significant for high pressure blowdown. Pressure increments are calculated and
when the pressure gradient becomes infinite the flow
is choked. If this occurs at the end of the pipe the
assumed flowrate is the converged choked flow solution.
If choked flow does not occur and the end of the line
is reached at the reservoir pressure, the non-choked
flow solution is obtained.

B. Defining the Reaction Kinetics and Component
Physical Properties. The rate expression needed for
use in a vent design model should represent the condition that would exist during the emergency. Kinetic
data based on the normal reaction rate are only useful
in cases when loss of heat transfer can be experienced.
A simple power law rate expression (usually first order)
will be sufficient if Arrhenius constants can be fitted.

For complex reactions, involving competing and un-
desirable side reactions, the most conservative approach
would be to size the vent system for the one or two

reactions that add the greatest amount of energy to the system over a given duration.

Use of thermal stability tests (DTA's) to determine the heat sensitivity of a given process mixture is desirable. Recent advances in analytical methods permit good calorimetric determination of heat of reaction. Heat of reaction data are critical for exothermic reactor vent sizing. Heat impact from fire is usually small in comparison, but should not be neglected.

Any convenient model for liquid phase activity coefficients can be used. In the absence of any data, the ideal solution model can permit adequate design. For multiple liquid phases (e.g. suspension processes) or increasing concentrations of polymers, some more realistic models are desirable (van Laar, Flory-Huggins, Wilson).

In design of emergency relief systems the intrinsic fluid properties can often make a difference. Usually a linear interpolation of density, viscosity (for Newtonian fluids) and heat capacity will provide suitable fluid properties, if the simulated temperatures fall within that range of data.

C. Will Two Phase Venting Occur? One of the key decisions in venting calculations is to determine whether two-phase vent flow will actually occur. Assume a reactor geometry as in Figure 3 with a vapor space and relief device located in the vapor space. One way for two-phase vent flow to occur is through gross entrainment of liquid with the discharging vapor. Another mechanism that can develop two-phase flow involves swelling or expansion of the contents due to bubble nucleation throughout the liquid volume. This fills the vapor space and the entire vessel with something approximating a homogeneous vapor-liquid mixture which will discharge as a froth. Before the onset of two-phase venting, there will be a brief period of all-vapor venting as illustrated in Figure 3.

Correlations are needed to predict whether two-phase flow will occur after vapor venting is initiated by rupture disc failure or relief valve opening. Research is needed in this area, but for the present we recommend the following correlations to predict batch swell. For systems with low viscosity (less than 500 cp) an equation based on bubble column hold-up is used to obtain a swell ratio:

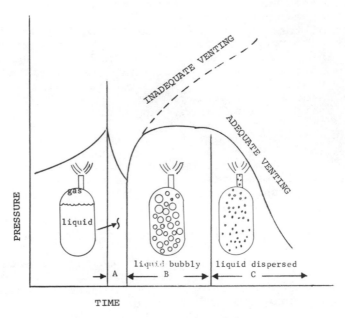

TIME

Figure 3.   Pressure variation with flow regime

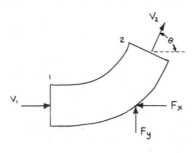

$$F_x = p_1 A_1 + p_2 A_2 \cos\theta + \frac{\rho Q}{g_c}\left(V_1 - V_2 \cos\theta\right)$$

$$F_y = \frac{Wg}{g_c} + \left(p_2 A_2 + \frac{\rho Q V_2}{g_c}\right) \sin\theta$$

Figure 4.   Forces on bends: (A) = Area ft²; (F) = Force lb_f; (g) = Acceleration of gravity ft/sec²; (g_c) = Conversion factor lb_m ft/lb_f sec²; (p) = Pressure lb_f/ft²; (Q) = Volume flow ft³/sec; (V) = Velocity ft/sec; (W) = Mass rate lb_m/sec; (ρ) = Density lb_m/ft³.

$$S = \frac{60 + 2V_s}{60 + V_s}$$

V_s is the superficial velocity of the gas in the reactor body in feet/minute. It conservatively assumes
all of the vapor is generated in the bottom of the
vessel.

For fluids with viscosity greater than 500cp, no
good general relationship is available. Experimental
work on one system allowed a swell ratio correlation
of the following form:

$$S = 1 + K R_V^{2/3} \mu^{1/2}$$

where K is a constant, R_V is the volume rate of gas
per liquid volume, and μ is the viscosity. When the
swell ratio exceeds the ratio of vessel volume to
liquid volume, two-phase homogeneous venting is
assumed.

IV. Mechanical Design

Specification of relief valves and rupture discs
must be done with care because of the potentially
tragic consequences of haphazard selection of size and
set or burst pressure. Disc burst conditions for example are very temperature sensitive and should be
selected for the temperature at which they will re-
lieve, not the normal operating temperature. Discs
also have a normal manufacturing tolerance of ±5% of
the set pressure. A 5% higher relieving pressure
could be significant in safely controlling a reaction.
Rupture discs are also susceptible to fatigue failure,
especially in pressure flucuating applications and
require periodic replacement. Relief valves have an
open area much smaller than their stated size and this
must be considered on selection, i.e. a 2" relief valve
may have an open area of 0.7 in^2.

Design of vessel and vent line pipe supports is
very important because very large forces can be encountered as soon as venting begins. Figure 4 shows
the equations and nomenclature to calculate forces on
pipe bends. The authors have heard of situations where
vent line bends have been straightened, lines broken
off, or vent catch tanks knocked off their foundations
by excessive forces. For bends, the transient effects
of the initial shock wave, the transition from vapor
flow to two-phase flow, and steady state conditions
should be considered. Transient conditions, however,
are likely to be so rapid as to not have enough dura-

tion to cause problems.

V.   Containment of Vented Material

Many of the materials handled are either explosive or toxic to people and the environment. Careful design is required to handle materials being vented.

Since much of the vented material will be liquid, separators such as knockout pots or tangential entry separators can provide disengagement and possible recovery. Figure 5 is a typical vapor-liquid separator design found to be effective for these applications. Inlet design superficial vapor velocity is about 100 ft/sec, with sufficient volume provided to accumulate the entire reactor liquid contents. The lip on the outlet vapor line and the horizontal plate to separate the accumulated liquid are important features to prevent re-entrainment.

Flammable or toxic vapors can be piped to a flare after separation of liquid is obtained. An important design problem in flare use is the very high vent rate experienced for a relatively short time, if an existing flare is used. Also back-pressure effects on the liquid separator vessel must be considered, especially if choked flow of vapor occurs downstream of the separator.

Another containment strategy for condensible or water-soluble emissions is to use a water quench system with the discharge being sparged into a large volume of cool liquid.

In very extreme cases, total containment can be provided to prevent any atmospheric emission or to provide a surge volume for controlled flaring, absorption, or other disposal methods. This approach, however, requires use of a very large pressure vessel to provide the required volume, and is usually only a last choice alternative.

VI. DIERS Program

While venting technology and methods are improving, considerable uncertainty remains as to the validity of various assumptions and accuracy of the correlations. Nearly all of the experimental data to verify calculations to-date are with air-water-steam systems.

Several chemical, refining, and engineering companies are currently in the process of forming a research institute to obtain realistic, verified design methods for reactor venting. The group is called DIERS (Design Institute for Emergency Relief Systems) and is sponsored by AIChE. Funds will be provided by the mem-

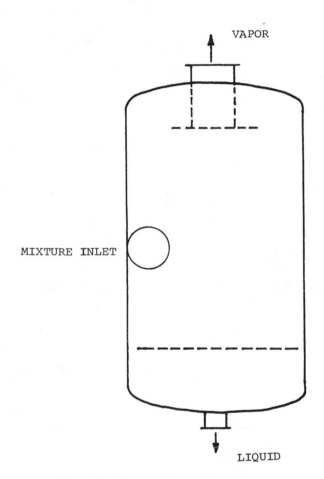

Figure 5.   Tangential vapor–liquor separator

ber companies on a schedule based on company size for
a four year program.

Emphasis of the program will be on 1) establishing
correlations for the batch swell in low and high viscosity systems verified by experiments, 2) establishing
good two-phase flow correlations verified by experiments
for vent piping and relief valves, with emphasis on
viscous, two-phase flow, 3) developing an overall two
phase venting design method, and 4) experimental verification of the design method on both small and large
scale with reacting and non-reacting chemical systems.

Membership is still available for companies interested in participating in this program.

VII. Summary

The reactor venting problem consists of several
key parts each of which must be understood and carefully handled: 1) the heat input either from exothermic reactions or other miscellaneous heat sources,
2) the batch swell mechanism, 3) the fluid mechanics
of the vent system, 4) the mechanical design of the
system, and 5) vent emissions control.

Literature Cited

1.   Boyle, W. J., Chem. Engr. Prog., (1967), 63,(8),
      61-66.

2.   Huff, J. E., Chem. Engr. Prog. Symp. Ser. - Loss
      Prevention, (1972), 7, 45-57.

3.   Lockhart, R. W. and Martinelli, R. C., Chem. Engr.
      Prog., (1949), 45,(1), 39-48.

4.   Martinelli, R. C., and Nelson, D. B., Trans. ASME,
      (1948), 70, 695-702.

5.   Hewlett, G. F. and Hall-Taylor, N. S., "Annular
      Two-Phase Flow", 23-27, Pergammon Press, Oxford
      GB, (1970).

6.   France, D. M. and Stein, R. P., Int. J. Heat and
      Mass Transfer, (1971), 14, 1407-1413.

RECEIVED January 18, 1979.

High Temperature Free-Radical Polymerizations in Viscous Systems

J. A. NORONHA, M. R. JUBA, H. M. LOW, and E. J. SCHIFFHAUER

Eastman Kodak Company, Rochester, NY 14650

Recently, we have been studying the runaway stages of some polymerization reactions. We are trying to learn more about designing equipment safely in the event a reaction gets out of control and runs away.

To do this we developed a computer model to predict the kinetic conditions during the runaway stage. The kinetic model is used to estimate the reaction rates, temperatures, pressures, viscosities, conversions, and other variables which influence reactor design.

To test our model, we set up small and large-scale tests for thermally-initiated polymerization of styrene.

The kinetic model predicted the observed reaction rates, pressures, rates of pressure rise and temperature rise within order-of-magnitude accuracies. The accuracy of the kinetic model was better for the large-scale tests.

We extended the kinetic model to other monomer systems such as styrene and methyl methacrylate. With these, we used common initiators such as benzoyl peroxide and azo-bis-isobutyronitrile. The results of these simulations compared closely with some published experiments.

With such modeling efforts, coupled with some small-scale tests, we can assess the hazards of a polymer reaction by knowing certain physical, chemical and reaction kinetic parameters.

Introduction

Several studies have been published to assess the kinetics of polymerization reactions at high temperatures. (1-7). However, most of these studies only describe experiments conducted at isothermal conditions. Only a few papers are based on adiabatic runaways (2). This paper is one of the first studies based on "first principles" characterizing adiabatic runaway reactions.

0-8412-0506-x/79/47-104-339$05.50/0

Discussion on Derivation of the Rate Equations

The polymerization rate equations are based on a classical free radical polymerization mechanism (i.e., initiation,
propagation, and termination of the polymer chains).
For thermally-initiated polymerization:

$$R_p = \left(A_p\right)\left(\frac{A_{dm}}{A'_t}\right)^{1/2} \left(\eta_{s,T}\right)^{1/2} (m)^2 \exp\left[\frac{E_t/2 - E_p - E_d/2}{(R)(T)}\right] \tag{1}$$

For a system employing a free radical initiator (i.e. a
peroxide or azo compound:

$$R_p = \left(A_p\right)(f)\frac{\left(A_{di}\right)}{\left(A'_t\right)}^{1/2} \left(\eta_{s,T}\right)^{1/2} (m) \left(I_2\right)^{1/2} \exp\left[\frac{E_t/2 - E_p - E_i/2}{(R)(T)}\right] \tag{2}$$

The following assumptions and theories are used in this
derivation:
1. For the thermally-initiated case, the initiation
rate has a second-order dependence on monomer concentration
as suggested by Flory[8] instead of a third-order dependence
as suggested by Hui and Hamielec[6].
When initiators are used, the initiation rate has
a first-order dependence on monomer concentration.
2. A quasi steady-state radical population exists.
3. The chain termination rate varies inversely with
the viscosity of the polymerization medium because of the
Trommsdorff Effect (i.e., the reduction of the macroradical
mobility with increasing reaction viscosity). This effect
significantly influences reaction rate[6,9,10].
4. The rate constants have an Arrhenius dependence on
temperature[11].
5. The solution viscosity is a function of the polymer
concentration and molecular weight, and can be determined by
the Hillyer and Leonard method[12].
6. The chain transfer reaction proposed by Hui and
Hamielic[6] and Olaj et al[13], affects the molecular weight
distribution but it does not affect the reaction rate.

Iterative Analysis

We started this study by developing a computer model to
predict the kinetic conditions during the runaway stage of a
reaction. The computer model is based on an iterative
analysis which permits a step-by-step computation of various
variables.

Figure 1 is a flow sheet showing some significant
aspects of the iterative analysis. The first step in the
program is to input data for about 50 physical, chemical and
kinetic properties of the reactants. Each loop of this
analysis is conducted at a specified solution temperature
T°K. Some of the variables computed in each loop are: the
monomer conversion, polymer concentration, monomer and
polymer volume fractions, effective polymer molecular weight,
cumulative number average molecular weight, cumulative
weight average molecular weight, solution viscosity, polymerization rate, ratio of polymerization rates between the
current and previous steps, the total pressure and the
partial pressures of the monomer, the solvent, and the
nitrogen.

Test Set-up

In order to test this computer model, we conducted
experiments on thermally initiated styrene polymerization in
sealed pressure vessels. We only measured pressures and
temperatures in these experiments. We conducted our tests
in two phases.
In Phase I (see Figure 2) we used a 300-cc stainless
steel pressure vessel, equipped with a 180-cc glass liner,
in which 100 cc could be polymerized. We used a pressure
gage, rated from 0 to 140 pounds per square inch. There
were 3 type J thermocouples - one in the center of the
solution, one in the reactor wall, and the third near the
heater outside the reactor. The experiments were conducted
in a high pressure bay and observed on closed circuit tele-
vision. The initial polymer concentrations of the test
reactants were either 0 or 15 or 30 percent by weight. An
electric heater controlled the ambient temperature of the
nitrogen - purged reactor, and supplied heat to initiate the
reaction.
Our computer model predicted the Phase I test results
with accuracy adequate for safety design even though there
were experimental errors. To reduce these experimental
errors, in Phase II, we made some equipment modifications
and used a larger reactor.
In Phase II (see Figure 3) we used a 2900-cc pressure
vessel, with a 2000-cc glass liner in which 1000 cc of
solution could be polymerized. This was a 10-fold increase
over Phase I. We used a pressure gauge similar to Phase I.
There were 5 type J thermocouples. Of these, there were 4
thermocouples within the reactor as compared to only 1 in
Phase I. Two were in the solution within the glass liner,
one was between the glass liner and reactor wall, and the

Define incremental monomer conversion

Physical, Chemical and Kinetic
Properties of the Reaction System

Starting Values:
 Concentrations of the monomer,
   solvent, polymer, and initiator
 System temperature
 Partial pressures of the monomer,
   solvent, and nitrogen
 Total pressure

Calculate
 Concentrations of the monomer,
   solvent, polymer, and initiator
 Solution viscosity
 Number average polymer mol. wt.
 Weight average polymer mol. wt.
 Polymerization Rate
 Reaction time
 Heat generated
 Heat losses
 Solution Temperature
 Partial pressures of the monomer,
   solvent, and nitrogen
 Total pressure
 Rate of pressure and temperature
   rise

(Monomer conc. in the next step)
=(Monomer conc. in the previous step)
-(Incremental monomer conversion)

Figure 1.   Iterative analysis

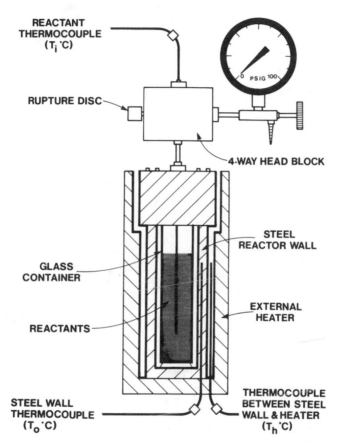

Figure 2.   Phase I test setup

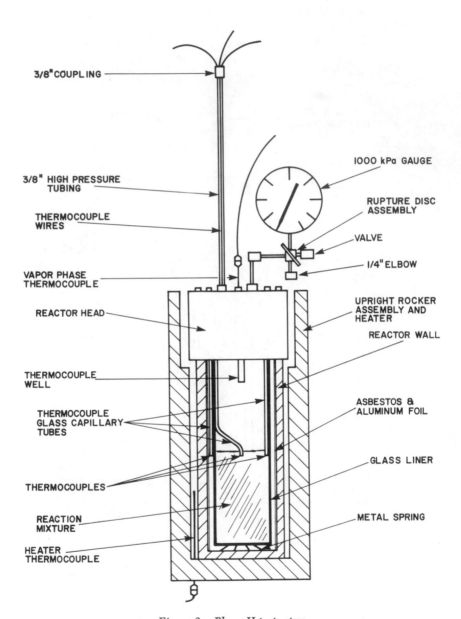

Figure 3.   Phase II test setup

fourth internal measurement was in the space above the
solution. The only external temperature measurement was
near the heater. We packed the space between the glass
liner and the reactor wall with asbestos. These Phase II
modifications made a big improvement over Phase I. (see
Table 1)

1. Since the solutions were not agitated in either
Phase I or Phase II, the temperatures were not uniform
throughout the solution. So in Phase II, the 3 additional
temperature sensors within the reactor gave us a better
estimate of the average solution temperature.

2. In Phase II the ratio of the reactor wall surface
to the reacting solution volume was six times lower. This
resulted in lower proportional heat losses which are difficult
to estimate. Hence, this resulted in lower computational
errors in Phase II.

3. The asbestos packing served two advantages; first,
it reduced heat losses and hence improved accuracy and
second, it replaced the vapor gap between the liner and
reactor wall. This minimized the convective heat transfer
of the vapor, which is also difficult to calculate.

Test Results

Since our model simulated the Phase II results more
accurately, we shall only discuss the Phase II results.
Let's discuss three tests in which the initial polystyrene
concentrations of the reactants were 0%, 15% and 30% by
weight respectively.

Figure 4 shows the observed pressure and temperature
data for Test 2. Initially, the external electric heater
controlled the system's temperature and supplied heat to
initiate the reaction. Later, as the reaction rate increased,
the reaction itself generated heat at a significantly higher
rate than the heater imput.

We estimated the average solution temperature as follows:

$$T_{av} = 0.3T_2 + 0.7T_3 \qquad (3)$$

The derivation was based on two assumptions. First, we
assumed a linear radial temperature gradient within the
solution. Second, we computed "T_{av}" at the radius at which
there were equal volumes of solutions on either side of it.

A common interpretation of the runaway stage is when
both the first and second derivatives of the average time-
temperature curve are positive. However, because we had an
external heat source in our tests, we had to account for the
external heater temperature "T_4".

TABLE I

COMPARISON OF PHASE I AND PHASE II TESTS

	Phase I Tests	Phase II Tests
Reactants Volume	100cc	1000cc
Surface/Volume Ratio	6:1	1:1
Temperature Measurements within Reactor	1	4
Solution Temperature Measurements	Less accurate	More accurate
Radial Heat Losses	More	Less
Radial Heat-Transfer Calculations	Less accurate	More accurate
Fit with Kinetic Model	Good	Better

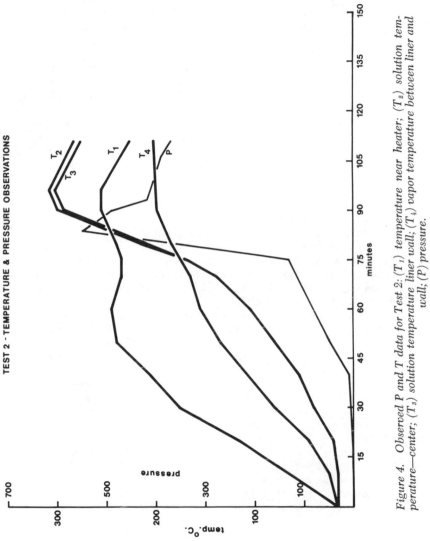

TEST 2 - TEMPERATURE & PRESSURE OBSERVATIONS

Figure 4. Observed P and T data for Test 2: (T_1) temperature near heater; (T_2) solution temperature—center; (T_3) solution temperature liner wall; (T_4) vapor temperature between liner and wall; (P) pressure.

We arbitrarily considered the _runaway stage_ to begin when the computed temperature difference between the T_4 and the average temperature of the solution goes through a minimum. For Test 1 (see Figure 5) this occurs when the average temperature was 100°C and T_4 was 150°C.

The _temperature variations_ within the solution were increased from Test 1 (in which the initial polystyrene concentration was 0%) to Test 2 (in which it was 15%) and to Test 3 (in which it was 30%) respectively. The maximum temperature differences between T_2 and T_3 were only 10° in Test 1, and 15° in Test 2 but 78°² in Test 3. The greater the temperature differences, the greater the error of calculating T_{av}. Hence, the computations for T_{av} were decreasingly accurate in Test 1, 2 and 3 respectively.

There's another reason why the computed solution average temperature had decreasing accuracies in Tests 1, 2 and 3 respectively. The reason is that we started with increasingly viscous solutions, which caused the response time of the temperature measurement to increase rapidly. This response time becomes even more significant because as the solution viscosity increases there are significant rises in the reaction rates and temperatures.

Now let's discuss the _pressure computations_. The observed reactor pressure is a sum of the partial pressures of nitrogen and the styrene monomer vapor. The vapor pressure of the styrene vapor is an increasing function of temperature and decreasing function of conversion. This is explained by the Flory-Huggins relationship (8).

Since we did not measure the conversion during the experiment, we computed the equilibrium vapor pressure at the average solution temperature. We believe that, for safety design, the equilibrium vapor pressure is an adequate estimate of the styrene vapor pressure. For example, even at a 50% conversion, the difference is only 10% at the experimental temperatures. Figures 6, 7 and 8 _compared_ the _observed pressures_ with the _computed total pressures_. The latter were based on the equilibrium vapor pressure. As expected, there were increasing variations in Tests 1, 2 and 3 respectively because of their higher initial conversions. From these figures we can verify that our pressure and temperature measurements were in phase with respect to time.

We next _estimated the conversions_ by using the observed pressures and temperatures and the Flory-Huggins relationship. Since the Flory-Huggins relationship is less accurate at higher conversions, we can expect these estimates of conversions to be of decreasing accuracy in Tests 1, 2 and 3 respectively.

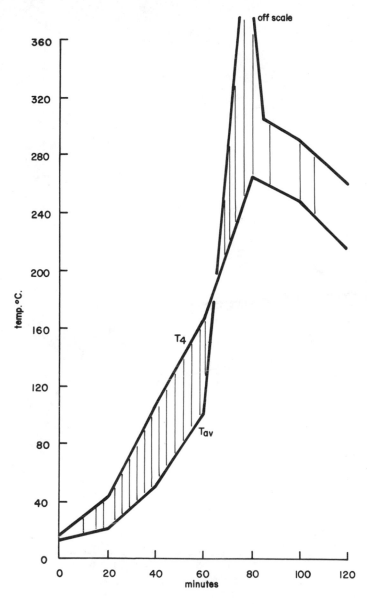

Figure 5. T_4 and T_{av} for Test 1

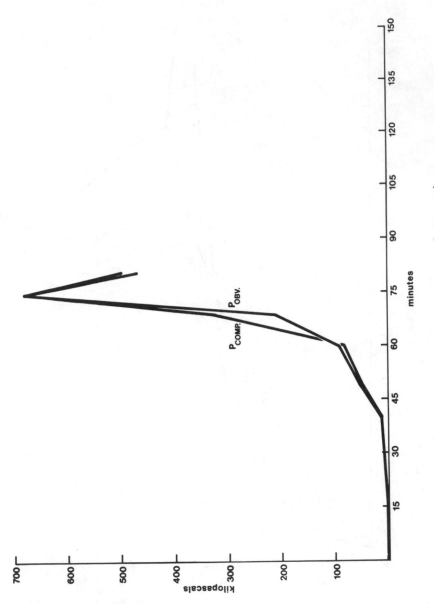

Figure 6.   Test 1 (observed and computed pressures)

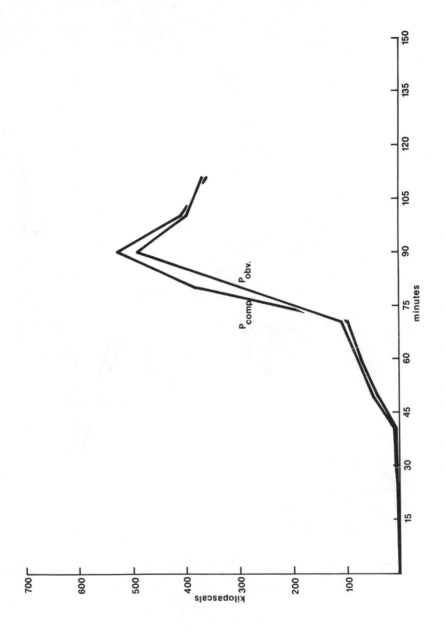

Figure 7.   Test 2 (observed and computed pressures)

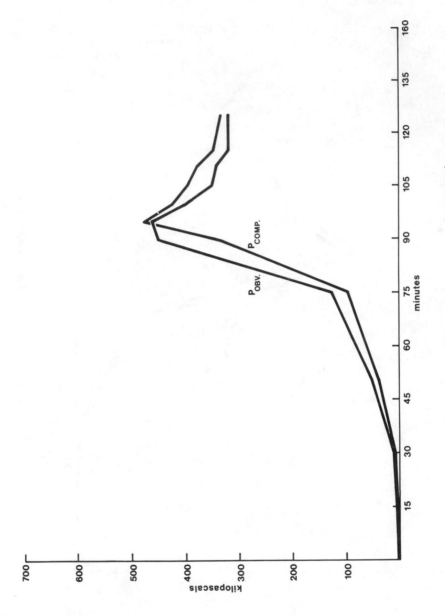

Figure 8.   Test 3 (observed and computed pressures)

Let's discuss the <u>reaction rate computations</u> based on the <u>kinetic model</u> with those derived from the <u>experiments</u>. At a given instant, these calculations are essentially "point" functions since they are independent of the path the reaction system has taken up to that given instant.

The kinetic model reaction rate is computed per equation (1) or equation (2) using the computed average solution temperature (T_{av}) and the estimated conversion(s).

The calculations for the experimental reaction rates are based on an unsteady state heat transfer analysis. We computed the overall heat transfer coefficient of the system and estimated the experimental rates as follows:

$$R_{exp} = f\ (\frac{dT_{av}}{dt},\ T_{av},\ T_4) \tag{4}$$

To simplify the equation (4) calculations during the runaway stage we drew the magnified plots of Test 1 during the 68 to 76 minutes (Figure 9) and for the 75 to 80 minute period (Figure 10).

We computed the percentage errors between the reaction rate computations based on the experiments with those based on the kinetic model. Note that, like the pressure and temperature comparisons, the accuracy of the calculations for reaction rates decreases as we compare Test 1 with Test 2 and Test 3. In Test 1 the error ranges from 3 to 21%, in Test 2 it was 10 to 21%, in Test 3 it ranged from 5 to 36%. In each test, the errors were in the lower order of its range during the earlier stages of the runaway reaction, and in the higher order of its range during the later stages.

We can explain why this decreasing accuracy occurs. The experimental reaction rate computations based on equation (4) are primarily functions of the computed average solution temperature (T_{av}). The kinetic model rate computations based on equation (1) or (2) are primarily functions of both "T_{av}" as well as the estimated conversion(s). Earlier we explained why we expected decreasing accuracies of estimating both the conversions and the average solution temperature in Tests 1, 2 and 3 respectively.

Other Monomer Systems - Comparison With Other Studies

The thermally-initiated styrene system is considerably simpler than most industrial applications. Though these experiments provided useful guidelines, it was difficult to develop broadly applicable design criteria without carefully evaluating a broad range of monomer, polymer and initiator systems. Hence we extended our kinetic model to some other monomer systems such as styrene and methyl methacrylate using common initiators such as benzoyl peroxide (BPO) and

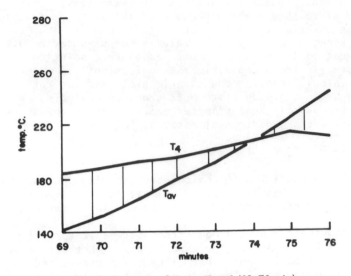

Figure 9. T_{av} and T_4 for Test 1 (68–76 min)

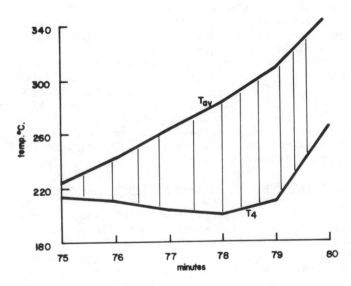

Figure 10. T_{av} and T_4 for Test 1 (75–80 min)

azo-bis-isobutyronitrile (AIBN). The results of these models compared quite favorably with some published experiments.

Most published studies relate only to isothermal experiments. Hence, in order to make such comparisons we modified our computations to assume isothermal conditions. Figure 11 compares our kinetic model with data by Hui and Hamielec (6) for styrene thermal polymerization at 140°C. Figure 12 compares out kinetic model with data by Balke and Hamielec (7) for MMA at 90°C using 0.3% AIBN. Figure 13 compares our kinetic model with data by Lee and Turner (5) for MMA at 70°C using 2% BPO. Our model compares quite favorably with these published experiments. The percent error was less than 5% in most of the ranges of conversions.

Limitations

1. The results of the model should be applied only to the runaway conditions of a system. They should not be applied to the non-runaway stage of the reaction.
2. The experiments were conducted at ambient temperatures up to 200°C. Hence, they do not relate to the high temperatures encountered if the reactor were exposed to an external fire.
3. The temperatures and pressures developed are a function of the heat transfer characteristics of the reaction system. Hence, our observed pressures and temperatures relate only to this particular system.

Conclusions

In conclusion, we have reviewed how our kinetic model did simulate the experiments for the thermally-initiated styrene polymerization. The results of our kinetic model compared closely with some published isothermal experiments on thermally-initiated styrene and on styrene and MMA using initiators. These experiments and other modeling efforts have provided us with useful guidelines in analyzing more complex systems. With such modeling efforts, we can assess the hazards of a polymer reaction system at various temperatures and initiator concentrations by knowing certain physical, chemical and kinetic parameters.

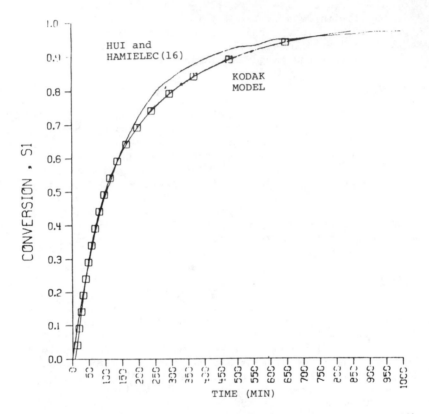

Figure 11.   Styrene thermal polymerization at 140°C, initial conversion −0%

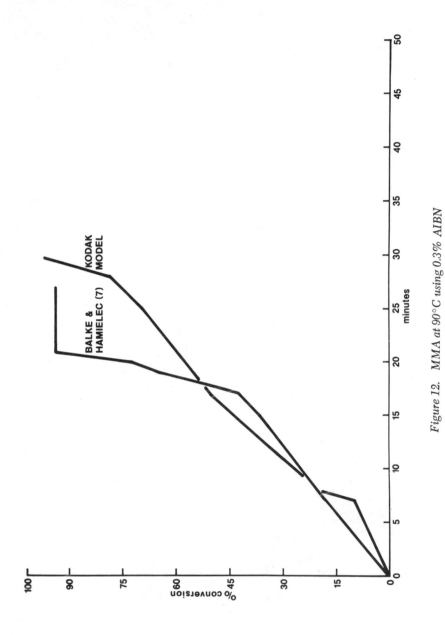

Figure 12.   MMA at 90°C using 0.3% AIBN

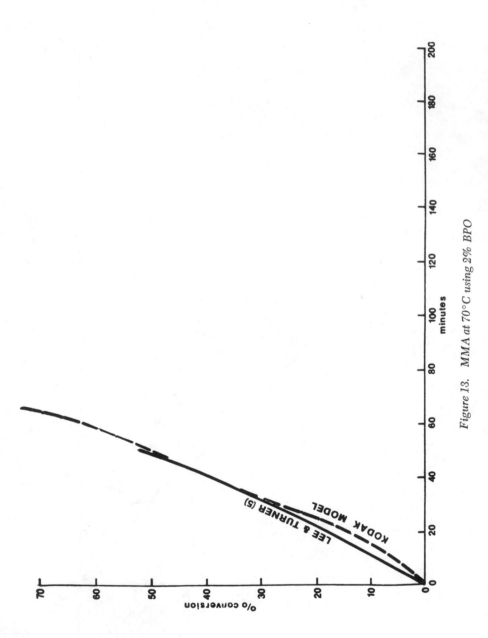

Figure 13. MMA at 70°C using 2% BPO

Glossary of Terms

$\eta_{S,T}$	$\equiv$	solution viscosity at conversion 'S' and temperature T°K, cp.
A_{di}	$=$	frequency factor for initiator initiation, 1/sec.
A_{dm}	$\equiv$	frequency factor for monomer thermal decomposition, liter/mole sec,
A_p	$\equiv$	Propagation frequency factor, liter/mole sec.
A'_t	$\equiv$	effective termination frequency factor, cp liter/mole sec,
E_d	$\equiv$	activation energy for monomer thermal decomposition, kcal/mole,
E_i	$=$	activation energy for initiation, kcal/mole.
E_p	$\equiv$	propagation activation energy, kcal/mole,
E_t	$\equiv$	termination activation energy, kcal/mole,
f	$\equiv$	initiator efficiency factor, dimensionless units.
$[I_2]$	$\equiv$	initiator concentration, mole/liter.
k_p	$\equiv$	propagation rate constant, liter/mole sec,
m	$\equiv$	monomer concentration, mole/liter.
P	$\equiv$	observed reactor pressure, kilopascals (gauge).
R	$\equiv$	Ideal Gas Law constant,
R_p	$\equiv$	polymerization rate, mole/liter sec.
s	$\equiv$	weight fraction of conversion, dimensionless units.
t	$\equiv$	time from start of experiment, minutes.
T_1	$\equiv$	temperature near heater (outside reactor), °C.
T_2	$\equiv$	temperature at center of glass liner (in the solution), °C.
T_3	$\equiv$	temperature at the inside wall of the glass liner (in the solution), °C.
T_4	$\equiv$	temperature between the glass liner and the reactor wall, °C.
T	$\equiv$	reaction temperature, !K.
T_{av}	$\equiv$	average solution temperature, °C.

Literature Cited

1. Sadawa, H., J. Polym. Sci., Polym. Lett. Ed., (1963),
    1, p. 305.

2. Sebastian, D. H. and Biesenberger, J. A., Kinetics and
    Thermal Runaway in Styrene Acrylonitrile Copolymerization -
    An Experimental Study. Presented at the 70th National
    AIChE Meeting held in Nov., 1977 in New York City.

3. Cardenas, J. N. and O'Driscoll, K.F., J. Polym. Sci., Polym.
    Chem. Ed. (1977), 15, p. 2097.

4. Barr, N.J., Bengough, W.I., Beveridge, G. and Park, G.B.,
    European Polym. J., (1977), 14, p. 245.

5. Lee, H.B. and Turner, D.T. Macromolecules, (1977), 10,
    (2), p. 226.

6. Hui, A. W. and Hamielec, A. E., J. Appl. Polym. Sci.,
    (1972), 16, p. 749.

7. Balke, S. T. and Hamielec, A. E., J. Appl. Polym. Sci.,
    (1973), 17, p. 905.

8. Flory, P. J., "Principles of Polymer Chemistry", p. 131,
    Cornell Univ. Press, Ithaca, N.Y., 8th printing, 1971.

9. Hayden, P. and Melville, H., J. Polym. Sci., (1960),
    43, p. 201.

10. Enal'ev, V.D. and Mel'nichenko, V.I., Mathematical
    Modeling of the Kinetics of Initiated Polymerization
    of Vinyl Monomers, U.S.S.R., Deposited Doc., Viniti,
    (1974), 319-74.

11. Odian, G., "Principles of Polymerization", p. 243,
    McGraw-Hill, N.Y., N.Y., 1970.

12. Hillyer, M. J. and Leonard, W. J., "Solvents Theory
    and Practice", R. W. Tess Ed., "Advances in Chemistry
    Series", p. 31, ACS, Washington, D.C., 1973.

13. Olaj, O.F., Kauffman, H. F., Breitenbach, J. W. and
    Bieringer, H., J. Polym. Sci., Polym. Lett. Ed. (1977),
    15, p. 229.

13. Olaj, O.F., Kauffman, H.F., Breitenbach, J. W. and
    Bieringer, H., J. Polym. Sci., Polym. Lett. Ed. (1977)
    2, p. 45.

RECEIVED March 15, 1979.

The Temperature Dependence of the Gel Effect in Free-Radical Vinyl Polymerization

K. F. O'DRISCOLL, J. M. DIONISIO, and H. KH. MAHABADI

Department of Chemical Engineering, University of Waterloo, Waterloo, Ontario, Canada N2L 3G1

In a series of papers ($\underline{1}$, $\underline{2}$, $\underline{3}$), a model has been developed and applied to describe free radical vinyl polymerization to moderately high conversion. Specifically, the model treats the case where polymerization rate increases with conversion as the reaction mixture becomes more viscous: the "gel effect". The model is based on the assumption that as a polymerization proceeds the increasing polymer concentration causes chain entanglement and thereby develops two populations of polymeric radicals: those that are smaller than some critical chain length, n_c and therefore mobile, and those radicals that are longer than n_c and therefore entangled and much less mobile. Consequently the termination reaction is described by two rate constants, k_t for reaction between two radicals of chain lengths less than n_c and k_{te} for radicals greater than n_c. For reaction between a chain of length less than n_c and one greater than n_c it was assumed that the termination rate constant is given by the geometric mean of k_t and k_{te}.

We now define a quantitative measure of the magnitude of the gel effect: the gel effect index, γ:

$$\gamma \equiv (R_p/R_{p,o}) - 1 \tag{1}$$

where R_p is the experimentally observed rate of polymerization at any given time and conversion and $R_{p,o}$ is the rate predicted by classical kinetics, which is to be expected at the same conversion and time in the hypothetical absence of a gel effect: i.e. with k_t unchanged.

When chain transfer is considered and following the derivation previously given ($\underline{1}$, $\underline{2}$) the instantaneous rate of conversion is then given by

$$\frac{dx}{dt} = A [I]^{0.5} (1 - x)(1 + \gamma) \tag{2}$$

where x is the fractional conversion of monomer to polymer and γ can be written in explicit terms of the model as:

$$\gamma = \frac{\tau(1 - \alpha)\exp(-\tau \cdot n_c)}{\tau_e - [C_m + (C_s[S] + C_I[I])/[M]](1 - \alpha)\exp(-\tau \cdot n_c)} \tag{1a}$$

The parameters τ and τ_e are reciprocal chain lengths and are given by expressions (3) and (4):

$$\tau = C_m + \{C_s[S] + C_I[I] + 2(Bf\ k_d[I])^{1/2}\}\ \frac{1}{[M]} \tag{3}$$

$$\tau_e = C_m + \{C_s[S] + C_I[I] + 2\alpha(Bf\ k_d[I])^{1/2}\}\ \frac{1}{[M]} \tag{4}$$

(see Nomenclature for a complete description of symbols).

It can be seen from equation (2) that when $\gamma = 0$ the model falls into the classical expression for the rate of conversion of free radical polymerization. Equation (1a) shows that this will be the case whenever all macroradicals have the same high mobility (i.e., as n_c tends to infinity) or when both entangled and non-entangled radicals have the same termination rate constant (i.e. α is equal to unity).

This model (1) has two adjustable, non-negative parameters, K_c and α. Having a correct definition of the onset of gel effect, (4) the value of K_c could, in principle, be given by the equation:

$$K_c = \phi_p\ \bar{X}_c^\beta \tag{5}$$

which is valid at that critical point, and the model could then describe an autoaccelerated polymerization reaction using only a single parameter, α_o which is a measure of the reduction in k_t caused by entanglement. In equation (5) ϕ_p is the volume fraction of polymer and $\bar{X}_c$ is the cumulative number average degree of polymerization of the polymer existing at the onset of gel effect.

It was found [1] that the values of K_c and α_o, obtained in minimizing the error of fitting experimental conversion-time data, satisfactorily described the temporal evolutions of the molecular weight averages. Also, the model performed better in the description of the experimental data when a value of $\beta = 1/2$ was used. We should point out that one of the postulates on which the kinetic equations were derived is that the rate constant of the termination reaction between entangled radicals, k_{t_e}, is proportional to the inverse first power of the entanglement density. While this postulate is certainly qualitatively correct, there is no a priori reason to have it take precisely this quantitative form.

In the work that follows, the experimental data were fitted by minimizing the sum of least squares and the differential equations were integrated numerically.

For each data set examined, the onset of the gel effect (which is the initial value for the integration of the differential equations) was taken at the point where there is a departure from linearity in the conversion-time plot. While a good argument can be made (4) for using another definition of the onset of the gel effect, the data available did not allow for a more detailed approach.

Temperature Dependence of the Model Parameters

Experimental conversion-time data, obtained from the literature, on the bulk free radical polymerization of MMA initiated by AIBN at several temperatures and initiator concentrations, were described by the model. However, the expressions for the rate of conversion and gel effect index were first simplified and rearranged.

Assuming that no chain transfer reaction was sufficiently important to be considered, useful simplifications result for equations (1a), (3) and (4), so that the gel effect index, γ, is given by the expression:

$$\gamma = (\frac{1}{\alpha} - 1)\exp\ (-\frac{n_c}{\nu}) \tag{6}$$

where $\underline{\nu}$ is the instantaneous chain length of the polymer produced from non-entangled radicals and is given by:

$$\nu = \frac{k_p\ [M]}{2\sqrt{k_t}\ f\ k_d[I]} \tag{7}$$

Also, the zeroth moment of the differential molecular weight distribution, DMWD, may be obtained by integration of the simplified equation:

$$\frac{d\lambda_o}{dt} = 2\ f\ k_d\ [I]$$

so that $\underline{\lambda_o}$ may be expressed as

$$\lambda_o = 2 \cdot f \cdot [I]_o \cdot [1 - \exp(-k_d \cdot t) \tag{8}$$

if the effect of shrinkage on the initiator concentration is neglected. Introducing equation (8) and the expression for the first moment of the DMWD, λ_1, into the equation

$$\bar{X}_n = \frac{\lambda_1}{\lambda_0}$$

the cumulative number average degree of polymerization, $\bar{X}_n$, may be expressed as:

$$\bar{X}_n = \frac{[M]_o}{2 \, f \, \varepsilon \, [I]_o} \cdot \frac{\ell n(1 + \varepsilon x)}{1 - \exp(-k_d \cdot t)} \tag{9}$$

where ε is the volumetric contraction coefficient. The parameter α in equation (6) has been expressed as (1):

$$\alpha = \alpha_o \left(\frac{K_c}{\phi_p \, \bar{X}_n^{1/2}}\right)^{1/2} \tag{10}$$

If equations (7) and (9) are then introduced into equation (6), as well as the correct expressions for n_c and the variation of ϕ_p with conversion, the expression for the gel effect index may be written as:

$$\gamma = (C_1 \, g_1 - 1) \exp(-C_2 g_2) \tag{11}$$

where:

$$g_1 = \left\{ \left(\frac{x}{1 + \varepsilon x}\right) \cdot \left[\frac{-\ell n(1 + \varepsilon x)}{f[I]_o (1 - \exp(-k_d \cdot t)}\right]^{0.5} \right\}^{0.5} \tag{12}$$

$$g_2 = \left(\frac{1 + \varepsilon x}{x}\right)^2 \cdot \left(\frac{1 + \varepsilon x}{1 - x}\right) (fk_d[I])^{0.5} \tag{13}$$

and

$$C_1 = \left[\frac{1 + \varepsilon}{\alpha_o^2 \, K_c} \left(\frac{[M]_o}{-2 \, \varepsilon}\right)^{0.5}\right]^{0.5} \tag{14}$$

$$C_2 = \frac{2 \, B^{1/2}}{[M]_o} \left(\frac{K_c}{1 + \varepsilon}\right) \tag{15}$$

Equation (2) can then be put into the form:

$$\frac{dx}{dt} = A[I]^{0.5}(1 - x)[1 + (C_1 g_1 - 1)\exp(-C_2 g_2)] \tag{16}$$

By minimizing the error of fitting experimental x vs. t data with equation (16) after the onset of gel effect, the parameters C_1 and C_2 can be obtained. Figures 1-4 compare the predictions by the model with experimental conversion-time data.

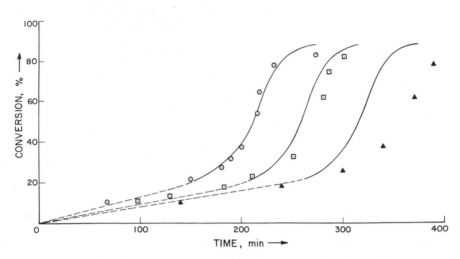

Figure 1.   Bulk polymerization of MMA at 45°C initiated by thermal decomposition of AIBN (5): (——) calculated from Equation 16. [I]$_o$ = 0.1 (○); 0.05 (□); 0.025 (▲).

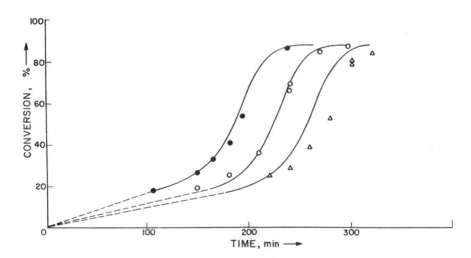

Figure 2.   Same as Figure 1 at 50°C: [I]$_o$ = 0.05 (●)[6]; 0.0277 (○)[7]; 0.0166 (△)[7]

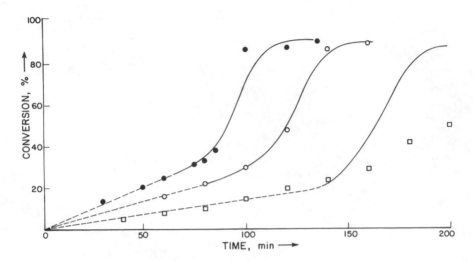

Figure 3. Same as Figure 1 at 60°C: $[I]_o = 0.05$ (●)[6]; 0.0183 (○)[8]; 0.0061 (□)[8]

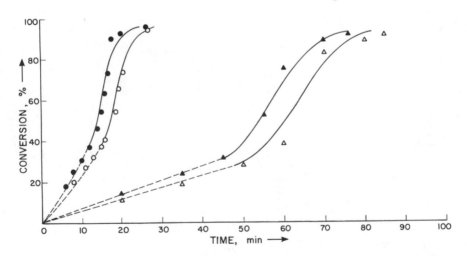

Figure 4. Same as Figure 1 at 70°C: $[I]_o = .3$ (△); 5 (▲) and at 90°C: $[I]_o = .3$ (○); .5 (●)(7)

It is important to note that C_1 and C_2 are quantitative
descriptors of the gel effect which depend only on the monomer,
temperature and reaction medium. The full description of γ,
given by equation (11), requires g_1 and g_2 which are functions of
the rate of initiation and extent of conversion. The kinetic
parameters used in these calculations and their sources are given
in Table 1. All data are in units of litres, moles and second.
Figure 5 shows the temperature dependencies of C_1 and C_2 and
Table 2 lists these and other parameters determined by fitting
the model to the data in Figures 1-4.

TABLE 1

Kinetic Parameters Used in Model
to Describe MMA Polymerization Initiated by AIBN

$T°K$	$[M]_o$	$A x 10^4$	B	$k_d x 10^5$	f	$-\epsilon$
318	9.16	0.87	54.9	0.0916	.453	.242
323	9.11	1.38	47.6	0.195	.465	.245
333	9.00	3.37	36.1	0.823	.498	.254
343	8.89	7.8	27.8	3.198	.529	.265
363	8.59	36.5	17.32	38.56	.598	.295
Source of data	(9)	Fig. 6	(7)	(13)	$\dfrac{A^2 B}{k_d}$	(7)

TABLE II

Kinetic Parameters Determined by Fitting Model
to Data of Figures 1 - 4

$T°K$	K_c	α_o	C_1 $(mol/\ell)^{1/4}$	C_2 $(sec)^{-1/2}$	$C_m x 10^4$
318	16.82	.0395	13.33	797	.112
323	15.50	.0525	10.38	638	.122
333	12.45	.0615	9.71	372	.230
343	10.30	.0775	8.30	233	.243
363	6.77	.0995	7.53	89	–

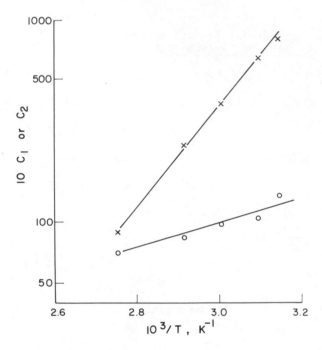

Figure 5. Variation of the model parameters C_1 (○) and C_2 (×) with tempera-
ture for the bulk free radical polymerization of MMA initiated by AIBN

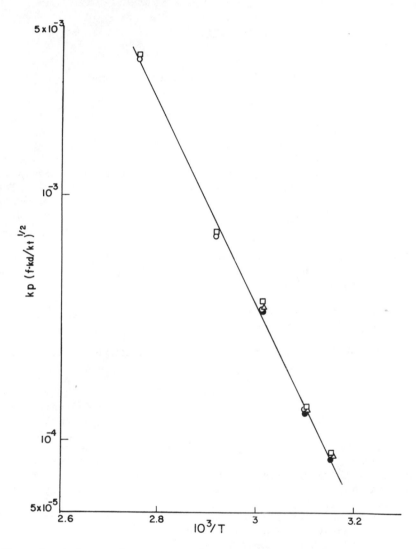

Figure 6.   Arrhenius plot of A for polymerization of MMA initiated by AIBN using data of Figures 1–4

The values of A were estimated from the time-conversion data before the onset of the gel effect. In this region $\gamma = 0$ so that integration of equation (2) yields

$$A = \frac{-\ln(1-x)}{[I]_o^{1/2} t} \tag{17}$$

The values of C_1 and C_2 in Table 2 have temperature dependencies given by

$$\ln C_1 = -1.69 + 1328/T$$

$$\ln C_2 = 11.05 + 5650/T$$

which may be used in conjunction with kinetic rate constants to describe MMA polymerizations over a range of temperatures. The apparent activation energies of these two parameters stem mostly from those of K_c and α_o. For K_c we calculate an apparent activation energy of -1.18 kcal/mol and for α_o of $+1.12$ kcal/mol. The small values of these numbers and the inadequacy of our present understanding of the "entanglement" phenomena suggest that we forgo discussion of them until we have more data or better theoretical understanding.

Model Description of the Bulk Polymerization of MMA with Chain Transfer to Monomer

Figures 1 - 4 show that when polymerizations were carried out at low concentrations of initiator and/or at low temperatures, the agreement between the model predictions and the experimental data is not so good. This is due to the fact that under those reaction conditions where R_p is low a large kinetic chain length is expected. When this is so, chain transfer to monomer becomes a reaction to be taken into account, since it markedly influences the chain length of the polymer being formed. A decrease in the instantaneous degree of polymerization, due to chain transfer to monomer, will reduce the concentration of the entangled radicals and, consequently, a decrease in the rate of polymerization is expected.

The general equation for the gel effect index, equation (1a) which incorporates chain transfer, was used in those cases where there was not a good agreement between model predictions and experimental data. The same values of K_c and α_o (derived from the values of C_1 and C_2 found at high rates) were used in the integration of equation (1) and the value of the constant of chain transfer to monomer, C_m, was taken as an adjustable parameter and used to minimize the error of fitting the time-conversion data by the model.

In all instances the expansion of the calculation to include

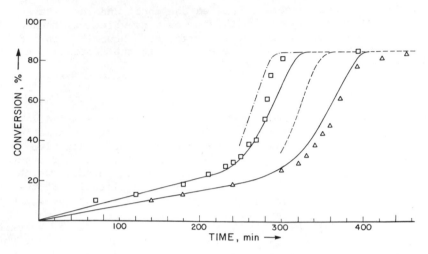

Figure 7. Same as Figure 1: (———) predicted with $C_m = .112 \times 10^{-4}$; (– – –) with $C_m = 0$

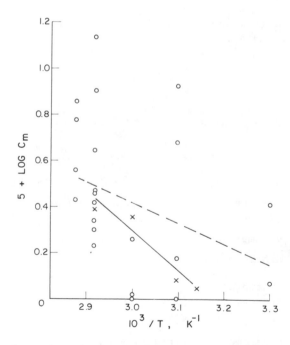

Figure 8. Comparison of values for chain transfer to MMA as tabulated ($\bigcirc$)[10] and found in this work ($\times$): (– – –) linear least squares for ($\bigcirc$); (———) for ($\times$)

C_m gave very good agreement between the model and the data.
Figure 7 shows an example of the improvement which was found.
The C_m values which minimized the error are shown in Figure 8
where they can be seen to compare well with the scattered values
tabulated (10) from the literature.

Solution Polymerization

The model was tested against solution polymerization data
for MMA reported by Schulz and Haborth (11). The minimization of
error in fitting the model to the data resulted in negative
values for α_o. This is physically unrealistic, and suggests that
the model needs modification. Further work is intended which will
refine the choice of initial condition for application of the
model and/or change the inverse dependency of k_{te} on entanglement
density to power greater than unity.

The ultimate goal is to bring together knowledge of solution
thermodynamics, rheology and polymerization kinetics so that the
latter, which is well described at low conversion (12,13), may
be better described in a continuous manner to complete conversion.

Acknowledgement

Support of this work by the National Research Council of
Canada is appreciated.

Greek symbols

α	dimensionless parameter defined as $(k_{t_e}/k_t)^{1/2}$
α_o	a lumped constant, equation (10)
β	exponent in equation (5); $\beta = 0.5$ in this work
γ	gel effect index; a measure of the amount of gel effect, defined by equation (1)
ε	volumetric contraction coefficient
λ_n	nth moment of DMWD
ν	instantaneous chain length of the polymer produced from non-entangled radicals
τ	reciprocal of ν, equation (3)
τ_e	a parameter defined by equation (4)
ϕ_p	volume fraction of polymer

NOMENCLATURE

A	a group of kinetic parameters, $A = k_p (f \, k_d/k_t)^{1/2}$
B	a group of kinetic parameters, $B = k_t/k_p^2$
C_I	chain transfer to initiator constant
C_m	chain transfer to monomer constant
C_s	chain transfer to solvent constant
C_1 and C_2	parameters defined by equations (14) and (15), respectively
f	initiator efficiency
g_1 and g_2	parameters defined by equations (12) and (13), respectively
$[I]$	initiator concentration
K_c	entanglement constant, equation (5)
k_d	initiator decomposition rate constant
k_p	propagation reaction rate constant
k_t	rate constant for the termination reaction between non-entangled radicals
k_{t_e}	rate constant for the termination reaction between entangled radicals
$[M]$	monomer concentration
n_c	critical radical chain length above which radicals are entangled
R_p	polymerization reaction rate observed at given conversion
$R_{p,o}$	polymerization reaction rate expected at given conversion in absence of gel effect
$[S]$	solvent conventration
t	time
x	conversion of monomer
$\bar{X}_n$	number average degree of polymerization
$\bar{X}_c$	number average degree of polymerization at the onset of gel effect

REFERENCES

1. Cardenas, J. and O'Driscoll, K. F., J. Polymer Sci. PCE, <u>14</u> 883 (1976)

2. Cardenas, J. and O'Driscoll, K. F., J. Polymer Sci. PCE, <u>15</u> 1883 (1977)

3. Cardenas, J. and O'Driscoll, K. F., J. Polymer Sci. PCE, <u>15</u> 2097 (1977).

4. Dionisio, J., Mohabadi, H. Kh., O'Driscoll, K. F., Abuin, E and Lissi, E. A., J. Polymer Sci. (in press)

5. Ito, K., J. Polymer Sci. PCE, <u>13</u>, 401 (1975)

6. Nishimura, N., J. Macrom. Chem., <u>1</u>, 257 (1966)

7. Balke, S. T. and Hamielec, A. E., J. Appl. Polymer Sci., <u>17</u>, 905 (1973)

8. Rokudai, M., Toyodkai, Y., and Saitou, Y., Nip. Kag. Kai, <u>8</u>, 1758 (1972).

9. Matheson, M. S., J. Am. Chem. Soc., <u>71</u>, 497 (1949)

10. <u>Polymer Handbook</u> (2nd ed.), Brandrup, J. and Immergut, E. H. Eds., Wiley, New York, 1975

11. Schulz, G. V., and Harborth, G., Makromol. Chem., <u>1</u>, 106 (1947)

12. Mahabadi, H. Kh. and O'Drixcoll, K. F. Macromolecules <u>10</u> 55 (1977)

13. Mahabadi, H. Kh., Ph. D. Thesis, University of Waterloo, 1976

RECEIVED January 12, 1979.

Reduction of Molecular Mobility Caused by Increasing Solution Viscosity

D. C. TIMM, C. HUANG, V. K. PALSETIA, and T. S. YU

Department of Chemical Engineering, University of Nebraska, Lincoln, NB 68588

The effect of media viscosity on polymerization rates and polymer properties is well known. Analysis of kinetic rate data generally is constrained to propagation rate constant invarient of media viscosity. The current research developes an experimental design that allows for the evaluation of viscosity dependence on uncoupled rate constants, including initiation, propagation and macromolecular association. The system styrene, toluene n-butyllithium is utilized. Steady state analysis explicitly evaluates model parameters. Dynamic simulations predict reactor start-up transients.

Historical Review

Kinetic Mechanism. The following ionic mechanism describes sytrene polymerization in a hydrocarbon solution with n-butyllithium as the initiator ([1-6]).

Initiator Association: $I_y \rightleftarrows y \ I$ $\qquad K_a$

Initiation: $I + M \rightarrow A_1$ $\qquad K_i$

Propagation: $A_j + M \rightarrow A_{j+1}$ $\qquad K_p \quad j = 1,2,3$

Polystyryl Anion Association: $A_j + A_k \rightleftarrows A_j *A_k \quad K_{eq}$

The initiator association number y depends upon temperature, solvent and concentration of the system and is reported to be between two and seven ([7]). The association of polystyryl anions results in species $A_j *A_k$ which do not react with styrene monomer. The macromolecular association is analogous to a reversible termination by combination.

0-8412-0506-x/79/47-104-375$05.00/0

Autoacceleration. Glass and Zutty (8) and Burnett
and Melville (9) reported an increase in the rate and
average degree of polymerization with increasing solution viscosity, heterogeneous conditions and chain
coiling for free radical, vinyl polymerizations. Autoacceleration is also called Trommsdorff (10) effect.
As the polymerization reaction proceeds, viscosity of
the system increases, retarding the translational and/
or segmental diffusion of propagating polymer radicals.
Bimolecular termination reactions subsequently become
diffusion controlled. A reduction in termination results in an increase in free radical population, thus
providing more sites for monomer incorporation. The
gel effect is assumed not to affect the propagation
rate constant since a macroradical can continue to
react with the smaller, more mobile monomer molecule.
Thus, an increase in the overall rate of polymerization
and average degree of polymerization results.

Constrained by the assumption that propagation
rates are independent of solution viscosity, termination rate constants have been correlated with media
viscosity, cumulative molar concentration of macroradicals, and molecular size (11,12,13).

Model Development. Rachow and Timm (14) derived
working relationships for the kinetic mechanism described. Degree of polymerization is considered to be
a continuous variable. For quenched samples a relationship correlating population density of associated polymer molecules as a function of time, degree
of polymerization and environmental factors is

$$\frac{\partial T(n,t)}{\partial t} + \frac{K_p M(t) A_{tot}(t)}{T_{tot}(t)} \frac{\partial T(n,t)}{\partial n} + \frac{T(n,t)}{\theta} = 0 \; ; \; n>1 \qquad (1)$$

A boundary condition is

$$T(1,t) = \frac{K_i T_{tot}(t) I(t)}{K_p A_{tot}(t)} \qquad (2)$$

The molar rate of change of polymeric species of degree
of polymerization n in a well-mixed, continuous flow
tank reactor is related to the kinetic rate of propagation of unassociated polystyryl anions plus their
withdrawal rate in the reactor's effluent. Feed
streams are void of polymeric substances, but contain
monomer initiator and solvent.

Styrene monomer concentration is described by

$$\frac{dM(t)}{dt} = \frac{M_{in} - M(t)}{\theta} - K_p A_{tot}(t) \, M(t) \tag{3}$$

Experimental initiator concentration is the sum of molar concentrations of associated as well as unassociated molecules of n-BuLi

$$\frac{dI_{exp}(t)}{dt} = \frac{I_{exp, \, in} - I_{exp}(t)}{\theta} - K_i' M(t) \, [I_{exp}(t)]^{1/y} \tag{4}$$

if initiator association is assumed to be at equilibrium.

The total, cumulative molar concentration of macromolecules is described by

$$\frac{dT_{tot}(t)}{dt} + T_{tot}(t)/\theta = K_i' M(t) \, [I_{exp}(t)]^{1/y} \tag{5}$$

The unassociated and associated polystyryl anion concentrations contribute to population density of the sample. Their molar concentration is

$$T(n,t) = [1 + K_{eq} A_{tot}(t)] \, A(n,t) \tag{6}$$

The integral molar concentration of unassociated macroanions is $A_{tot}(t)$. The total cummulative molar concentrations of all polymeric species at time t is described by

$$T_{tot}(t) = [1 + K_{eq} A_{tot}(t)] \, A_{tot}(t) \tag{7}$$

The polymerization system for which experiments were performed is represented by the mathematical model consisting of Equations 1 and 7. Their steady state solutions are utilized for kinetic evaluation of rate constants. Dynamic simulations incorporate viscosity dependency.

Kinetic Evaluation

Experiments were performed in an isothermal, well-mixed, continuous tank reactor. Uncoupled kinetic parameters were evaluated as follows from steady state observations.

Population Density Distribution. Integration of Equation 1 yields a semilogarithmic relationship.

$$\ln T(n,ss) - \ln T(1,ss) = - \frac{T_{tot}(ss)}{\theta K_p A_{tot}(ss) M(ss)} (n-1) \qquad (8)$$

Values of population density $T(n,ss)$ at various degrees of polymerization n are obtained through gel permeation chromatography (GPC) analysis. Experimental analysis yields numerical values for the slope and intercept.

$$(-slope) = \frac{T_{tot}(ss)}{\theta K_p A_{tot}(ss) M(ss)} \qquad (9)$$

$$intercept = T(1,ss) \qquad (10)$$

Rearrangement of Equation (9) results in Equation (11) from which the product $K_p A_{tot}(ss)$ can be evaluated

$$K_p A_{tot}(ss) = \frac{T_{tot}(ss)}{(-slope)\,\theta M(ss)} \qquad (11)$$

The quantity $K_p A_{tot}(ss)$ can independently be obtained from conservation of styrene.

$$K_p A_{tot}(ss) = \frac{M_{in} - M(ss)}{\theta M(ss)} \qquad (12)$$

Equations (11) and (12) provide two experimental methods for evaluation $K_p A_{tot}(ss)$. Experimental agreement confirms the accuracy of the population density distribution obtained by GPC.

Simulataneous solution of Relationships (7 and 9) results in the working relationship

$$\theta M(ss)(-slope) = \frac{1}{K_p} + \frac{K_{eq}}{K_p^2} \frac{T_{tot}(ss)}{\theta M(ss)(-slope)} \qquad (13)$$

If a set of isothermal data is obtained at various levels of viscosity, a regression analysis will allow for the evaluation of the two rate constants as functions of viscosity.

Initiator Association. Experimental initiator concentration is the concentration of associated and unassociated n-BuLi. If initiator is predominantly associated and at equilibrium, Equation (2) may be expressed as

$$\frac{T(1,ss)K_p A_{tot}(ss)}{T_{tot}(ss)} = K_i' [I_{exp}(ss)]^{1/y} \qquad (14)$$

where $K_i' = K_i [K_a/y]^{1/y}$

The kinetic mechanism subject to quenched samples requires that each polymer molecule formed will contain one butyl residue

$$I_{exp}(ss) = T_{tot}(ss) - I_{exp,in} \qquad (15)$$

The value of initiator association and the rate constant may be evaluated. Viscosity is not expected to have a significant cage effect as in free radical systems, but the extent of association may be dependent on viscosity, or other properties of the fluid media.

Polymerization Dynamics

Numerical simulations of reactor start-up were programmed, predicting monomer and initiator concentrations, total polymer concentration, weight and number average molecular weights, viscosity and population density distribution dynamics. The following two relationships obtained from steady state observations were utilized in the simulation.

$$\mu(t) = \mu_0 = 2.059 \times 10^{-10} (W_{tot}(t))^{3.874} (T_{tot}(t))^{-1.125} \qquad (16)$$

$$K_{eq}(t) = 1.16 \times 10^{-19} \exp(+26340/RT) \mu^{-.2025} \qquad (17)$$

Viscosity of monomer feed solution is μ_0

Moment Analysis. The zeroth moment is the molar concentration of polymer and is expressed by Equation 5. The first moment is proportioned to the mass of polymer formed and is related to monomer concentration Equation 3. The second moment WA(t) is expressed by

$$\frac{dWA(t)}{dt} + \frac{WA(t)}{\theta} = \frac{2K_p M(t) W_{tot}(t)}{[1+K_{eq}A_{tot}(t)]} \qquad (18)$$

The second moment is used to evaluate weight average molecular weight.

Reactor start-up simulations require initial values of A_{tot}, I_{exp}, $T(n,t)$ and T_{tot} be zero. Monomer concentration is non-zero. The boundary condition, Equation 2, is expressed as

$$T(1,t) = \frac{K_i^! \, [I_{exp}(t)]^{1/y}}{K_p} \, [1+K_{eq}(t)A_{tot}(t)] \qquad (19)$$

Population Density Response Surface. The algorithm method of characteristic is used to reduce the partial differential Equation (1) into a set of coupled ordinary differential equations. Since $T(n,t)$ is an exact differential, then

$$\frac{\partial T(n,t)}{\partial t} \frac{dt}{ds} + \frac{\partial T(n,t)}{\partial n} \frac{dn}{ds} = \frac{dT(n,t)}{ds} \qquad (20)$$

Comparison of Equations (1) and (20) yields the following ordinary differential equation set.

$$\frac{dt}{ds} = 1$$

$$\frac{dn}{ds} = \frac{K_p M(t) A_{tot}(t)}{T_{tot}(t)}$$

$$\frac{dT(n,t)}{ds} = \frac{T(n,t)}{\theta}$$

Integrating the three equations with respect to time yields

$$t - t^* = s$$
$$n(t^*,n^*,t)-n^* = \int_o^t \frac{K_p M(t) A_{tot}(t)}{T_{tot}(t)} \, dt - \int_o^{t^*} \frac{K_p M(t) A_{tot}(t)}{T_{tot}(t)} \, dt \qquad (21)$$

$$T(n,t) = T(n^*,t^*)\exp[(t^*-t)/\theta] \qquad (22)$$

Constants of integration are t^* and n^*. For practical applications, initial conditions specify that $n^* > 0$, $t^* = 0$ and boundary conditions require $n^* = 1$, $t^* > 0$. If $t^* = 0$, $n^* = 1$, a ground curve passing through the origin can be generated. This function $n(0,1,t)$ was evaluated through Runge-Kutta-Gill integration. Values of population density along this ground curve are evaluated using Equation (22) and the boundary condition $T(1,0)$.

To evaluate a specific molar concentration $T(n_1,t_1)$ the point $[n_1,t_1]$ is initially located. If it lies above the principal ground curve, i.e., $n_1 > n(0,1,t_1)$, it is necessary that the ground curve passing through the point $[n_1, t_1]$ originates from the initial condition plane and $t* = 0$, $n* > 1$. Equation (21) may be arranged such that $n* = n_1 - n(0,1,t_1)$. Equation (22) coupled with the null initial condition $T(n*,0) = 0$ yields a zero population density $T(n_1,t_1) = 0$. Sufficient time has not elapsed for the formation of this size macromolecule.

If the point $[n_1, t_1]$ lies below the principal ground curve $n(0,1,t_1)$, the ground curve passing through $[n_1, t_1]$ must originate from the boundary condition plane, $t* > 0$, $n* = 1$. To implicitly evaluate the constant $t*$, this ground curve is generated by the translation

$$n(0,1,t_1) - n_1 = n(0,1,t*) - 1$$

The implicit constant $t*$ is evaluated from the principal ground curve. An interpolation of the function $n(0,1,t)$ yields the constant of integration $t*$ when $n(0,1,t) = n(0,1,t*)$. The boundary condition $T(1,t*)$ coupled with Relationship (22) yields the population density of polymeric species of size n_1 at time t_1. If the principal ground curve passes through the point $[n_1, t_1]$, then $t* = 0$, $n* = 1$.

A principal advantage for the above formulation is the reduction of integrations required. Along a ground curve, population densities are uncoupled from nearest neighbors. Thus a combination of two integrations (Equations 23 and 24) plus variable coefficients and linear translations allows for the explicit evaluation of the molar concentration of any polymeric specie. The classic solution requires an ordinary differential equation at each degree of polymerization plus variable coefficients. Nearest neighbors are coupled. Arguments of integration are simpler functions when the method of characteristics is applied.

Experimental

Equipment. The reactor was 1.523 liter, 316 stainless steel cylindrical, jacketed vessel equipped with two multiblade, paddle-type agitators. Tracer studies showed the reactor was well-mixed. A thermocouple measured temperature and was recorded continuously. Feed tanks, tubing, pumps and valves were made of stainless steel and had teflon seals.

Procedure. Concentration of n-BuLi in the feed was measured by titration (15). The reactor was filled completely with styrene monomer solution in toluene initially. Time was measured from the moment initiator and monomer feed was initiated. The reaction was

allowed to continue for six to seven residence times.
Reactor samples were quenched and analyzed for styrene
concentration; polymeric weight was obtained gravimetrically from dried samples.

Gel Permeation Chromatography. The instrument
used for GPC analysis was a Waters Associates Model
ALC-201 gel permeation chromatograph equipped with a
R401 differential refractometer. For population
density determination, polystyrene powder was dissolved
in tetrahydrofuran (THF), 75 mg of polystyrene to 50
ml THF. Three μ-styragel columns of $10^2, 10^3, 10^4$ Å
were used. Effluent flow rate was set at 2.2 ml/min.
Total cumulative molar concentration and population
density distribution of polymeric species were obtained
from the observed chromatogram using the computer program developed by Timm and Rachow (16).

Steady State Population Density Distributions.
Representative experimental population density distributions are presented by Figure 1 for two different
levels of media viscosity. An excellent degree of
theoretical (Equation 8) / experimental correlation
is observed. Inasmuch as the slope of population
density distribution at a specific degree of polymerization is proportional to the rate of propagation for
that size macroanion, propagation rates are also
observed to be independent of molecular weight.

Uncoupled Rate Constants. An initial evaluation
of polymerization kinetics is presented in Figure (2),
constrained by viscosity invariant rate constants K_p.
The slopes of these straight lines give initial
estimates of K_{eq}/K_p^2 according to Equation (14). Figure 3 presents graphically a power law relationship
between K_{eq}/K_p^2 and viscosity at 21°C and at 16.6°C.
More scatter in Yu's data may be attributed to the
use of an older GPC instrument of relatively low resolution. The ratio K_{eq}/K_p^2 is temperature-sensitive;
a change of the order of five times is observed if the
temperature is reduced by 4.4°C and viscosity is kept
constant.
    Using this preliminary observation a comprehensive
analysis of data will allow for the elucidation of
the viscosity dependency. If K_p and K_{eq} are assumed
to be power functions of viscosity with an Arrhenius
temperature coefficient

$$K_p = a \exp (-E_p/RT) \mu^b$$

$$K_{eq} = c \exp (E_{eq}/RT) \mu^d$$

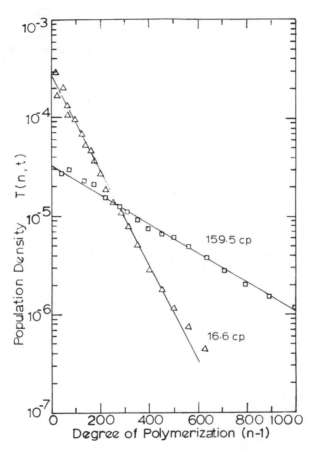

Figure 1.   Steady state population density distributions

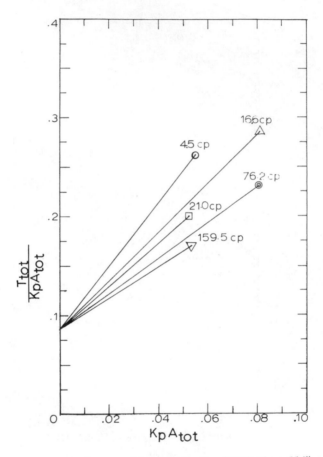

Figure 2.   Preliminary kinetic evaluation (21°C, $K_p = 11.5$)

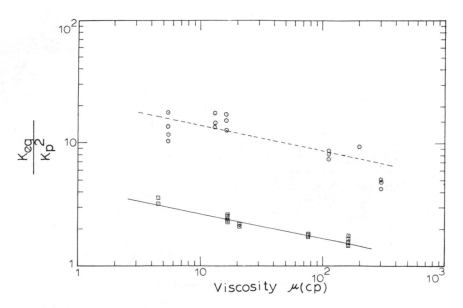

Figure 3. *Propagation and polystyryl anion association kinetics:* (– – –) *21°C;*
(——) *16.6°C.*

The activation energy of the propagation reaction (E_p)
and that of association equilibrium reaction (E_{eq}) are
reported to be 6.13 Kcal/gmole and 38.6 Kcal/gmole respectively (17). A non-linear search of the data
(Equation 14) will define the constants a,b,c, and d.
Data at 16.6°C and 21°C were incorporated with a least
square objective function using Luus and Jaakola's (18)
method. The analysis resulted in the following relationships:

$$K_p = 4.44 \times 10^5 \exp(-6130/RT) \, \mu^{-0.0002}$$
$$K_{eq} = 6.77 \times 10^{-38} \exp(+50860/RT) \, \mu^{-0.2025} \, K_p^2 \qquad (32)$$

This shows that K_p is independent of viscosity. Equilibrium association of polystyryl anions, is dependent
on solution viscosity.

Initiation analysis is presented by Figure 4. A
power curve fit of the data yields values of y and
$K_i^!$ to be 3.571 and 0.002137 respectively. The data
scatter may be attributed to the fact that concentration of primary ions T(1,ss) is very sensitive to
chromatogram heights. Contributions of molecules in
the low molecular weight tail of a chromatogram are
significant to the total molar concentration, which is
subject to a high degree of experimental uncertainty.
This error is further magnified in reading a semilogarithmic population density distribution. Timm
and Kubicek (19) report a value of ȳ to be 3. Thus,
the current value is of similar magnitude. Current
results were obtained using GPC columns with plate
counts in excess of 1,000 plates/ft. The cited research utilized equipment of the order of 100 plates/ft.

Polymerization Dynamics. Relationships presented
were utilized for the simulation of monomer concentration, number and weight average molecular weights,
and population density distributions for two experimental observations. Experimental values of these
variables are in reasonable proximity of calculated
values.

Monomer concentration dynamics are presented in
Figure 5. Additional observations for Run 5 are
accurately correlated during the reactor startup and
at final steady state. The observation at one residence time, Run 4, may be in error. The total cummulative, molar concentrations of macromolecules as a
function of time are presented in Figure 6. The
errors associated with this dependent variable are also
evident during the steady state analysis of initiation

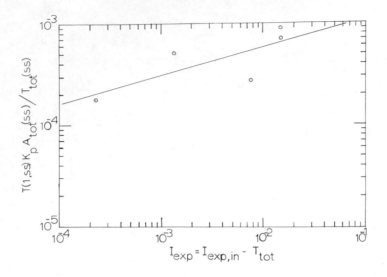

Figure 4. Initiation kinetics at 21°C

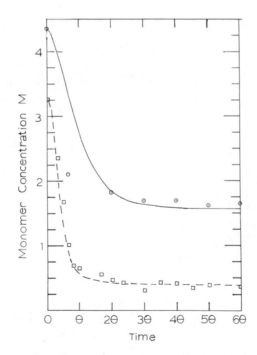

Figure 5. Monomer dynamics at 21°C: (– – –) Run 4 simulation data; (——) Run 5 simulation data.

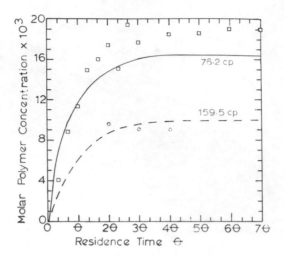

Figure 6.   *Cumulative molar concentration dynamics:* (– – –) *Run 4 simulation data;* (———) *Run 5 simulation data.*

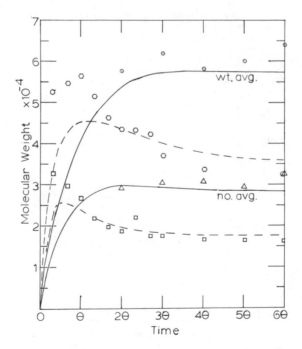

Figure 7.   *Molecular weight dynamics:* (– – –) *Run 4 simulation data;* (———) *Run 5 simulation data.*

kinetics as seen in Figure 4. Smoothing of data by
a power function can result in substantial error for
a given observation. The dynamic character of the
response is present, but a final error at steady state
is prevalent, but is within experimental measurement.

Number and weight average molecular weight transients are summarized in Figure 7. The more viscous
conditions of Run 4 resulted in an overdamped response;
whereas, the less viscous conditions of Run 5 resulted
in overshoot. The simulation was more damped and delayed than the experimental response.

A simulation of population density dynamics is
presented in Figure 8. The development of macromolecules is evident. With increasing sample time, the
distribution approaches the final, steady state exponential distribution. The wave within the distribution describes a local maximum in population density as
a function of time and degree of polymerization and is
a direct result of reaction dynamics during reactor
start-up. The nature of the chain growth polymerization mechanism is significantly different than that
associated with a typical free readical polymerization.
The dimeric polystyryl anion association stabilizes
the reactive site. Thus, significant time is required
to achieve high molecular weights. If the polystyryl
anion complex is considered a terminated species,
analogies exist with step growth polymerization mechanisms where an ionic intermediate reacts to form a
bond and then catalytically deactivates. Examples
include polyesters and polyamides. Yet the chemical
nature of the polystyryl complex is distinctively
different that that of a deactivated molecule.

Conclusions

Viscosity has negligible effects on the propagation rate constant, but an appreciable effect on the
equilibrium polystyryl anion association. The propagation rate is independent of molecular size.

Simulation dynamics are verified experimentally.
Model simulation is useful for developing control
systems, improving operation and predicting dangerous
regions for polymerizations. An example may be a runaway temperature due to loss of heat transfer. Modeling may result in greater reactor efficiency allowing
increased conversion with products of uniform speci-
fications or materials presently non-existent but
desirable.

The current study was limited by viscosity due to
mechanical limitation in reactor design. An experiment

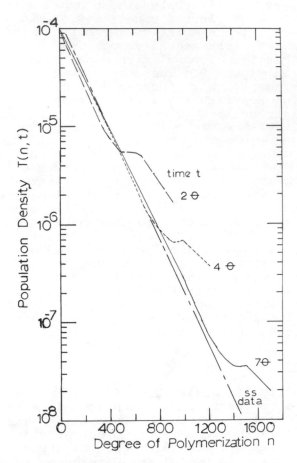

Figure 8. Population density distribution dynamics: Run 5

to obtain a higher viscosity of an increased average molecular weight was aborted due to mixing power and bearing limitations.

Nomenclature

A_j Molar concentration of polystyryl anion of size j

A_j*A_k Associated polystyryl anion complex

A_{tot} Total cumulative molar concentration of polystyryl anions

E_p Activation energy of propagation reaction in the Arrhenius equation

eq   Activation energy of polystyryl anion association equilibrium reaction in the Arrhenius equation

I   Initiator concentration

I_{exp} Experimentally measured initiator concentration

$I_{exp,in}$ Experimentally measured initiator concentration in the initiator feed solution

I_y Associated initiator concentration

K_a Initiator association equilibrium constant

K_{eq} Polystyryl anion association equilibrium rate constant

K_i Initiation rate constant

K_i' Effective initiation rate constant $(=K_i \, K_a/y \ ^{1/y})$

K_p Propagation rate constant

M   Monomer concentration in the reactor

M_{in} Monomer concentration in the monomer feed solution

n   Degree of polymerization

R   Gas law constant Kcal/gmole °K

ss   Steady state

t   Time

T   Absolute temperature

T(n,t)   Molar concentration of polymeric species of degree of polymerization n at time t

T_{tot} Total, cumulative molar concentration of polymeric species

y Association number for the initiator

θ Residence time

μ Viscosity

Abstract

A kinetic study for the polymerization of styrene, initiated with n-BuLi, was designed to explore the Trommsdorff effect on rate constants of initiation and propagation and polystyryl anion association. Initiator association, initiation rate and propagation rates are essentially independent of solution viscosity. Polystyryl anion association is dependent on media viscosity. Temperature dependency correlates as an Arrhenius relationship. Observations were restricted to viscosities less than 200 centipoise. Population density distribution analysis indicates that rate constants are also independent of degree of polymerization, which is consistent with Flory's principle of equal reactivity.

Dynamic simulations for an isothermal, continuous, well-mixed tank reactor start-up were compared to experimental moments of the polymer distribution, reactant concentrations, population density distributions and media viscosity. The model devloped from steady-state data correlates with experimental, transient observations. Initially the reactor was void of initiator and polymer.

Acknowledgements

The authors wish to express their appreciation to the Engineering Research Center of the University of Nebraska-Lincoln for financial support of this research.

Literature Cited

1. Welch, F. J., J. Amer. Chem. Soc., (1959) 81, 1345.
2. O'Driscoll, K. F., Tobolsky, A. V., J. Poly. Sci., (1959) 35, 259.
3. Morton, M., Fetters, L. J., and Bostick, E. E. J. Poly. Sci. (1963) C1 311

4. Morton, M., Bostick, E. E., and Livingni, R., Rub. Plast. Age (1961) 42 397

5. Hsieh, H. L., J. Poly. Sci., (1965) A3 153.

6. Worsfold, D. J., and Bywater, S., Can. J. Chem. (1960) 38, 1891.

7. Lenz, R. W., "Organic Chemistry of Synthetic High Polymers", 412, Interscience, New York, (1967).

8. Glass, J. E., and Zutty, N. L., J. Poly. Sci. (1966) 4 (Al) 1223.

9. Burnett, G. M., and Melville, H. W., Proc. Roy. Soc. (London) (1947) 89 (Al) 494.

10. Trommsdorff, E., Kohle, H., and Legally, P., Makromol, Chem. (1947) 1 169.

11. Burkhart, R. D., J. Poly. Soc., (1965) A3 883.

12. Vaughan, M. F., Trans. Faraday Soc. (1952) 48 576.

13. Hui, A. W. and Hamielec, A. E., J. Poly Sci. (1968) C25 167.

14. Timm, D. C., and Rachow, J. W., Adv. Chem. Series (1974) 133 122.

15. Gilman, H. and Houbein, G., J. Amer. Chem. Soc., (1944) 66 1515.

16. Timm D. C., and Rachow, J. W., J. Poly. Sci. (1975) 13 1401.

17. Ries, M. J., and Kubicek, L. F., AIChE Student Bulletin (1974) 15 26.

18. Luus, R., and Jaakola, T. H. I., AIChE J. (1973) 19 760.

19. Timm, D. C., and Kubicek, L. F., Chem. Engr. Sci. (1974) 29 2145.

RECEIVED February 1, 1979.

INDEX

INDEX

397

S